Edwin Dudley Smith

The New Method

An Interesting Thesis on General Debility and Nervous Exhaustion

Edwin Dudley Smith

The New Method
An Interesting Thesis on General Debility and Nervous Exhaustion

ISBN/EAN: 9783744670111

Printed in Europe, USA, Canada, Australia, Japan

Cover: Foto ©berggeist007 / pixelio.de

More available books at **www.hansebooks.com**

THE NEW METHOD.

AN INTERESTING THESIS ON

GENERAL DEBILITY

AND

Nervous Exhaustion;

WITH A FULL DESCRIPTION OF THE

GENERATIVE ORGANS OF MAN,

AND THEIR PECULIAR DISEASES.

ALSO,

THE NEW METHOD OF TREATING NERVOUS EXHAUSTION.

By EDWIN D. SMITH, A.M., M.D.,

Professor of the Principles and Practice of Surgery, Diseases of the Genito-Urinary System, and Clinical Surgery; Professor of Pathological Anatomy and Histology, Diseases of the Nervous System, and Clinical Medicine; Attending Physician at the Clinics for Venereal and Skin Diseases; Operating Surgeon for Diseases of Genito-Urinary System of Women; Author of Medical Works on Special Diseases of the Genito-Urinary Organs of both Male and Female, etc.

OFFICE HOURS—8 A.M. TO 1 P.M., AND 5 TO 8 P.M.

OFFICE:
100 East Twenty-ninth Street, cor. Fourth Avenue,
NEW-YORK.

TABLE OF CONTENTS.

INTRODUCTION.

SOME years ago it occurred to me that it would be beneficial to mankind to issue a small and popular work, which would be acceptable to all without as much as possible using the many technical terms so prolific with writers on medical subjects, that would be plain and intelligent to the many readers and to those suffering from the maltreatment, neglect, and ignorance of self-styled physicians; but circumstances prevented the consummation of my desires until the present moment, when I determined to issue this work, that it may serve as a "*beacon light*" to warn off the many sufferers from the rocky coast that would lead them to destruction through the so-called doctors that attempt to foist their pretensions of skill on a too-willing and deluded public; and should this little work, with its plain and unvarnished truths, lead and instruct any sufferers to a true and intelligent account of their ailments, I shall be well rewarded for my time and labor. It is with this object in view, and for the true benefit of mankind, that I shall endeavor to

demonstrate clearly, and succinctly, the damaging results that may occur from neglect of those warning
symptoms, that disease in its many forms shows itself
to the practiced eye of a physician, and to point them
out and impress them upon the mind of my readers,
that they may be warned thereby and so seek proper
and skillful medical aid.

With the above object in view, and with a sincere
desire to help and aid our fellow-man in his hour of
distress and despondency—those hours when all the
world seems dark and forlorn to his excited imagina-.
tion from the result of his over-indulgence by forcing
nature on too early to do the work of adult life before those organs God has endowed man with, that he
may "increase and multiply after his kind," have at-
tained their full development—and man rendered incapable of procreation, and giving everlasting shame,
misery, and unhappiness to the human race.

Now, we have been endowed with certain procreative organs so delicate and elaborate in their minute
anatomy, that secrete an important fluid—the *liquor
seminalis*, or seminal liquid—the richest and most
elaborate and complex of all the secretions of the
human body; truly the very essence and foundation
of life that strengthens the body, invigorates the
mind, renders the nervous system powerful, that we
may exercise our memory and judgment for our future happiness, and those of others near and dear to

us. So, if we waste this precious fluid, so important in all its relations to our well-being, in a manner never intended or ordained by nature, but too common among most young men of the present day, we sap the foundation of our being, overwhelm the nervous system in its powerful functions, and then must suffer the many disorders always resulting from this over-indulgence, and causing the nervous system to deteriorate, the brain and all the important organs to become impaired, and giving us as an unwelcome legacy—melancholy, impotency, nervousness, and a general decay of all the faculties. And when the sufferer becomes in this condition, and realizes, perhaps too late, the wretchedness of his situation, and he is no longer able to enjoy the society he may be so well fitted to adorn; that he is incapable of sexual intercourse; by form a man, but not in truth; without the healthy power of mind and body—and so becoming morbidly affected with distrust of his fellow-man, extremely sensitive, and perhaps leading on to acute mania or melancholia.

The nervous system is very important in its power over life, and on all human happiness has a most direct bearing; therefore its uses, and the danger of being overtaxed by any cause, should be fully appreciated by the public; and if by constant mental work, or an over-indulgence in any of the passions, the nervous system becomes impaired, it should receive prompt and immediate attention by seeking the advice of one who has given these peculiar disorders his time and especial study for many years, and from whose experience and practice can discover the cause, and as promptly apply the proper means for a speedy and permanent cure ; so that the nervous power becoming exhausted, the brain disordered, and the system at fault generally, they should be corrected, and a new impulse given to life by the application of those methods, new to the practice of medicine, that I have perfected and brought to the acme of science and skill ; and from my large experience, I am able to command success, and the heartfelt thanks of the many patients who have committed their cases to my care and judgment.

There are many cases, also, in which the cause of many troubles are due to malformation, arrest of development, and of congenital origin ; and that can only be relieved by means of a surgeon's skill. To all such, we would advise a prompt consultation, in which they may be assured of the utmost consideration in the skillful treatment of their case.

Many of these cases that require surgical interference have their direct influence on the happiness of married life ; and as the perfection of offspring and rightful succession of estates depends on the physical perfection of those about to enter the married state, it should not be consummated while there is any

doubt that through weakness, malformation, or disease its object would be useless, and the contracting
parties rendered miserable for life, when, by a careful
perusal of this instructive work, and a consultation with
one who has made this department of medicine his
special study, may afford the certain relief and cure.

And to the many who, through the example of
designing or indiscreet persons, have given the passions free sway in many ways, and so have impaired
their health, feel that they are in danger of premature
decay; let him, ere it is too late, look for those means
that are now at his disposal, and so gain back his
wasted strength and vigor, and feel that he is " once
again a man;" and that he may continue on to a
ripe old age and happy end, surrounded by a bright
and happy family.

How many, without number, of both sexes, may
be laboring under *mental troubles* and *physical peculiarities* connected with their generative systems, and
which are the greatest importance of their lives; and
yet it is only their own secret, locked in their innermost thought, and fearing to speak about it to their
most intimate friends, because they will not be understood or their motives appreciated; and so they
conceal their secret, and carry it on a heavy burden
to the grave, a constant bane to their lives, and perhaps cause an early death, when by a little boldness,
and a proper trust and confidence in their physician,

they might be relieved of so much worry and unhappiness, and live to enjoy old age and honor.

Therefore, in the following pages of this work, we will as briefly as possible give a description, first of the anatomy and physiology of the generative organs, that they may be readily understood, and their general importance in the human economy fully appreciated, and we will then endeavor to give to the intelligent reader the causes, variety, symptoms, etc., of the diseases and peculiarities that may affect the generative organs, that may cause them wholly or in part to fail in those important functions for which they were ordained; and upon a proper discharge of which, depends and rests the entire happiness in the future of many individuals and families, as well as the prosperity of the country: for it is an unquestioned fact, that according to the perfection and vigor of these organs and their healthy action, depend, in a great measure, the health and perfection of the offspring of man, as is taught to us in our daily experience.

In all respects we wish it distinctly understood that we stand second to none; that we will always keep in the advance of practical medicine, that our patients may have the benefits of the progress of the times, and the advantages of all new remedies and new methods that may be advanced, and will be faithfully tried and proved; and I would call particular attention to my *new method*, approved by the faculty in the treatment of Spermatorrhœa, or seminal weakness, and which will be fully spoken of in our article on that pernicious bane, to most young men of the present day, and in all cases of any discovery, I do not adopt it or employ it in my practice, until fully tried, calmly and without prejudice, in its chemical and therapeutical effects, and then finally testing it by intelligent experiments on proper cases; by so doing, we keep step with the advance in physics, and have at our command all the best and most approved remedies and instruments for the relief of many of those depressing "ills that flesh is heir to;" nor shall I spare any expense, to obtain all those remedies and appliances that may be of benefit to my many patients.

This is one of the many reasons that has induced me to issue this work at the present time, that I may bring before my readers a full and clear account of their troubles, at the same time offer them a means of speedy and complete cure, and so I have considered it my duty, through this book, to bring the results of experience and practice before an intelligent public, both for them and our mutual advantage.

As it would be interesting and instructive to our readers, I insert also some of the most peculiar and striking cases from my case-book, as they may be useful and give a correct notion to many in regard to their complaints, and to show the treatment they have received that all future patients may see that their troubles are fully understood and appreciated ; that they, too, with many others, may look for speedy relief and sympathy. It will also be noticed that all these cases are so arranged that the author and patient may not be known to any one personally, though many of my patients, feeling truly grateful for the relief afforded at my office, have given me permission to refer all those who may apply personally to me.

In conclusion I would say, after careful perusal of this work, that the human organs of generation, in their deep and complex nature, require the most careful and cautious treatment, and as this has been my special study for the past years, and the treatment of all cases of nervous debility, venereal infection, loss of sexual power, malformations, and all complaints arising from a disorganization of the reproductive organs, whether constitutional or acquired, will be faithfully carried out in all its details, and that all those who may apply to me for advice or assistance may be assured of not only intelligent treatment, but the most inviolable secrecy, sympathy, and skillful attention.

E. D. SMITH, M.D.,

Physician and Surgeon,

No. 100 East 29th st., cor. 4th ave.,

New York City.

CHAPTER I.

ANATOMY AND PHYSIOLOGY OF THE GENERATIVE ORGANS.

THESE organs, so very important in their relations to the happiness of mankind and for the proper procreation of his species, are so varied and complex in their nature and minute anatomy, that in a work of this kind we can not give a full account, but will give their anatomical and physiological relations in such a manner that the intelligent reader may fully understand them, and so be better able to judge of the nature and seriousness of his malady; that he may realize how important all these structures are in keeping the mind and body in a healthy condition, that it may perform the many functions and conditions it is so eminently fitted for.

This work is only intended for the consideration of special maladies, with their causes, prevention, and treatment, and all those depending upon the organs of generation in the human species, in which a proper performance of their functions, with which they are endowed, has ever been considered so necessary and essential to our health and well-being, both mental and physical. They have always excited the

(13)

admiration of all anatomists, physiologists, and microscopists in their beautiful arrangements and complex nature; so perfect in all their many forms, and so well adapted by nature to perform that work she has set for them to do; and, when we realize the delicate structure—so minute that it is not visible to the naked eye—and their peculiar fitness for the functions required of them, we understand how easily they may be impaired and their utility destroyed by any abuse or maltreatment that they may be subjected to through our passions and desires.

In our classification of the organs of generation we would divide them into—first, those lying within the abdominal cavity; second, those lying within the pelvis; and third, those lying external to the body. Of the first we have the kidneys with a portion of the ureters, and the blood-vessels supplying them, and the nerves controlling their action; and in the second, or those within the pelvis, we have the urinary portions of the urethra, the tubes coming from the testicles that convey the *liquor seminalis*, or seminal fluid containing the procreative seed, the seminal vesicles, the prostate gland, bladder, and the important portions of the urethra; and in the third, or those lying external to the body, the penis and urethra, with the scrotum, testicles, and appendages (Fig. 1).

The KIDNEYS are two true glandular organs, solely intended for the secretion of the urine, and are situ-

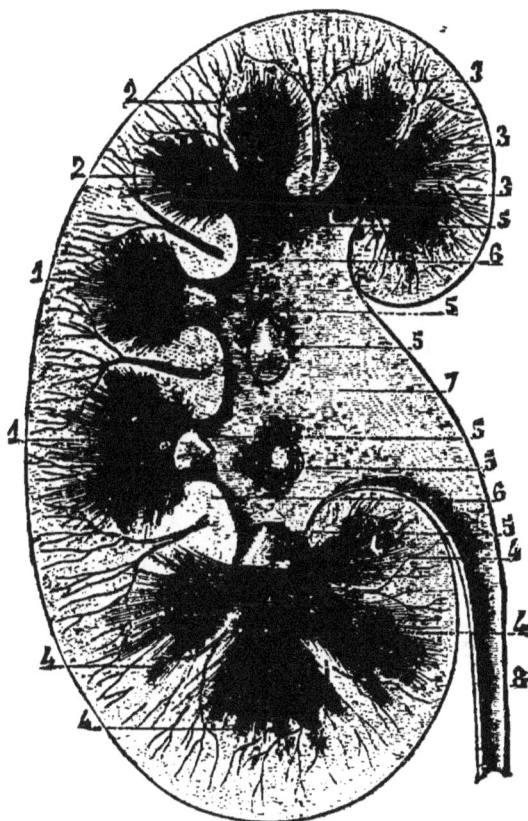

FIG. 1.—VERTICAL SECTION OF THE KIDNEY.—FLINT.

1, 1, ¨, 2, 3, 3, 3, 4, 4, 4, 4, pyramids of malpighi; 5, 5, 5, 5, 5, 5, apices of the pyramids surrounded by the calices; 6, 6, columns of Bertin; 7, pelvis of the kidney; 8, upper extremity of the ureter.

ated in the back part of the abdominal cavity, behind
the peritoneum; one lying in each lumbar region, on
each side of the spinal column, and extending downward from the eleventh rib to the crest of the ilium
—the right being somewhat lower than the left on
account of the liver being above it; the left having
the spleen above. The kidneys are capped by two
small ductless glands—the *suprarenal capsules*—and
are retained in their position by the vessels which enter them, and are surrounded by cellular tissue and
fat. Each one is about four inches in length, and in
shape resembles the kidney-bean. They are invested
by a fibrous capsule, thin and smooth and easily
removed from the surface. On section the kidneys
present two portions to us—the cortical and the pyramids — both consisting of blood-vessels and the
urinary tubes that lead down to the hilum, and
carry the urine, after its abstraction from the blood,
to the pelvis, and from that to the ureters. Now, as
the kidneys secrete the urine from the rich supply of
blood passing in by the renal artery, we can readily
see how these important organs may be affected and
influenced by the passions—as instance the effect of
suddenly increasing the flow of urine, and a desire
to void it from the effect of fear on infants and animals, and also in those patients laboring under the
effects of organic structure, with difficulty in passing
the urine. The mind worried and harassed from this

complaint will cause great increase in the secretion of this fluid, and so greatly add to his already existing irritation by the frequent calls to empty the bladder, and might be very serious from retention of urine.

The *Ureters*, these tubes lined by mucous membrane, commence at the pelvis of the kidney, and convey the urine to the bladder, and are called the " excretory duct of the kidney ;" they are about the size of a goose-quill, and from sixteen to eighteen inches long, and they pass down, one on each side of the body, to the back and lower'part of the bladder, and enter that organ by passing obliquely between its muscular and mucous coats, for about an inch, and terminate by a constricted orifice, thereby preventing regurgitation of urine.

The Bladder (see Fig. 2) is a muscular membranous sac, situated in the pelvis at the lowest part of the body, behind the pubes and in front of the rectum in the male, and the uterus in the female ; this organ is the reservoir of the urine, and admits of great distension, so much so that when fully distended it rises out of the pelvis into the abdominal cavity ; it is oval in shape, and usually contains about a pint of urine, and has four coats, a serous, muscular, cellular, and mucous, named from without inwards ; the muscular coat acting by its contractions to expel the urine, and is held in its position by the various ligaments that connect the bladder to the different por-

2

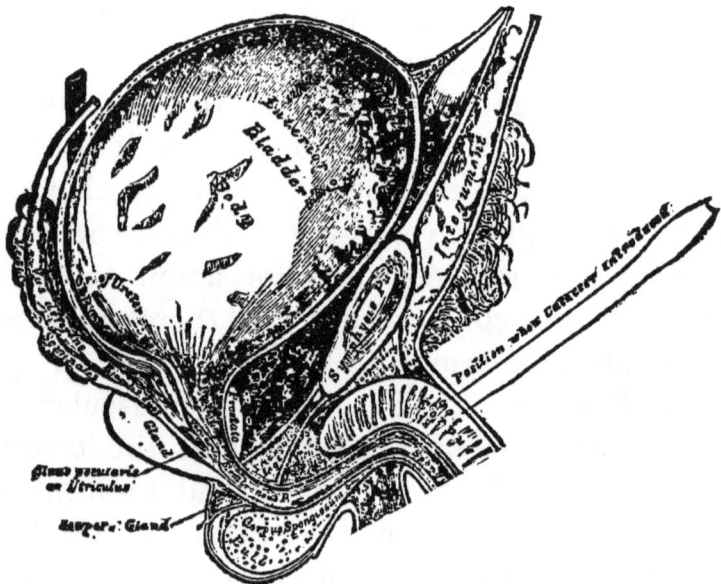

FIG. 2.—VERTICAL SECTION OF BLADDER, PENIS, AND URETHRA.

tions of the pelvis. Upon the inner surface of the base or fundus of the bladder, and behind the orifice of the urethra, is the *trigone vesicale.*

The Urethra commences at the apex of the *trigone* at the neck of the bladder, and extends the entire length of the penis, to the meates urinarius, and presents a double curve in the flaccid state ; its. length is about eight to nine inches, and consists of three portions, the prostatic, membranous, and spongy portions; the prostatic portions pass through the prostate gland, and upon its floor we have the *verumontanum* or *caput gallinaginis,* an elevation of the mucous membrane lining the canal, and serves to prevent the semen passing back to the bladder, on the floor of which on each side are the *prostatic sinuses* containing numerous openings, the orifices of the ducts of the prostate gland, and in front of which are the mouths of the ejaculatory ducts, and the *seat of trouble* in seminal emissions ; the urethra is very vascular, has some elasticity, and is lined by mucous membrane, very thin, and without any muscular fibers, and its lower part surrounded by muscles, the *accelerator urinæ,* and *compresses urethra* to assist in expelling the urine.

We have two glands opening into the urethra: first, the prostate gland at the mouth of the bladder, and in shape and size somewhat like a horse-chestnut ; it is a pale, firm body surrounding the neck of the bladder, and consists of three lobes—two lateral, and a middle lobe—and secretes the prostatic fluid—a milky fluid of acid reaction, consisting of columnar epithelium and granular nuclei—and placed beneath and on each side of the membranous portion of the urethra, we have Compers glands about the size of peas, with their excretory ducts opening in the urethra.

The Penis (see Fig. 2), "the organ of copulation," consists of the cavernous portions or bodies (corpora cavernosa), and the spongy portion or body (corpus spongiosum), this latter extending the whole length of the penis, from its bulb to its termination in the glands which overlap the anterior portion or ends of the corpora cavernosa. These bodies are covered by integument or skin, very thin, and continuous with the mucous membrane, which covers the glands of the penis, and contains numerous small and highly-sensitive papillæ, slightly raised, and supposed to be the seat of pleasure or of pain in the parts ; and upon the glands we have numerous small, lenticular, sebaceous glands, that secrete a matter of very peculiar odor, becoming easily decomposed.

We will now turn our attention to the anatomy of those parts—the scrotum with its contents, the testicles and appendages, as these are the most important organs concerned solely in the secretion of the most important fluid—the liquor seminalis. The scrotum

or purse is a loose bag of skin just below the root of
the penis, and divided into two lateral halves by a me-
dian line or raphe; the left is somewhat longer than
the right, due to greater length of the spermatic cord
on that side, and consists of the integument and dar-
tos; the latter is a thin
layer of connective tissue
inclosing the testes, and
divides the scrotum into
two cavities, forming in
the center the *septum
scroti.*

The Testicles, those
most important organs of
the human economy, are
two glands having an ex-
cretory duct, and are situated in the scrotum, one
in each of its cavities, and
their secretion consists of
the true semen, the pro-
creating fluid which en-
dows the ovum or human
egg with its vital power;
they are suspended by

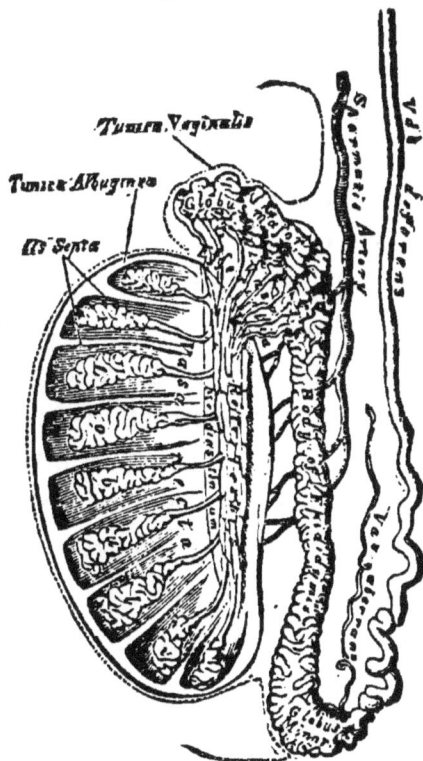

FIG. 3.—Vertical Section of the Testicle,
to show the arrangement of the Ducts.

the spermatic cord, which consists of the excretory
duct or vas-deferens, vessels, and nerves. During
early fœtal life they lie in the abdominal cavity, be-

hind the peritoneum, and before birth they descend
along the inguinal canal, and emerging at the external
abdominal ring, pass down to the scrotum, carrying
with it the spermatic cord and vessels, surrounded by
a fold of peritoneum, which forms the serous covering
of the testes, and its upper portion becoming obliter-
ated, forms a distinct sac.   They are supplied by the
spermatic arteries which arise from the main arterial
trunk, the aorta.   Each testicle is of an oval form,
flattened laterally, and on each outer edge lies the
epididymis, consisting of a body (Fig. 3), head (globus major), and tail (globus minor), and they are invested with the testicle with tunics or coats (the tunica vaginalis, tunica albuginea, and tunica vasculosa),
the first being most important; we have spoken of it
above.   The minute anatomy of these important organs is too complex to give one a perfect idea how vast
and wonderful they are, as when we consider that each
testicle consists of three or four hundred lobules, each
of which is a long, coiled tube, said to be several feet
in length, and only visible under the microscope.   It is
in these minute tubes that the semen is secreted from
the rich supply of blood passing over them ; and they
therefore serve for the secretion and elimination of
the seed, as it passes along these minute tubes to the
head of the epididymis, and in the vas-deferens or
seminal duct, and by that tube is conveyed to the
vesicular seminalis or seminal vesicles (Fig. 4) situated

at the base, and underneath the bladder, and serves as a storehouse for the semen until required for use, when it passes out through the ejaculatory duct to empty into the urethra.

I have dealt thus far with the anatomy of these

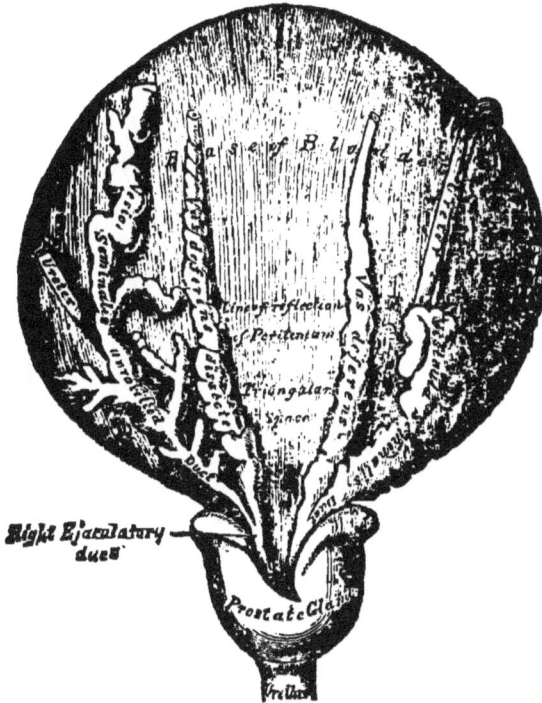

FIG. 4.—Base of the Bladder, with the Vasa Deferentia and Vesiculæ Seminales.

organs that my readers may fully appreciate their importance, and to show that being so minute in their structure, and elaborate in their formation, that they can not be forced on in their functions by self-abuse, without their most complete disorganization.

This important fluid, the secretion from these glands, in its physiological structure, consists of the true semen, a thick, whitish fluid, having a peculiar odor, and the liquor seminalis, with solid particles, the seminal granules and spermatozoa; this liquor seminalis being a transparent, colorless fluid, of an albuminous composition, and the seminal granules round,

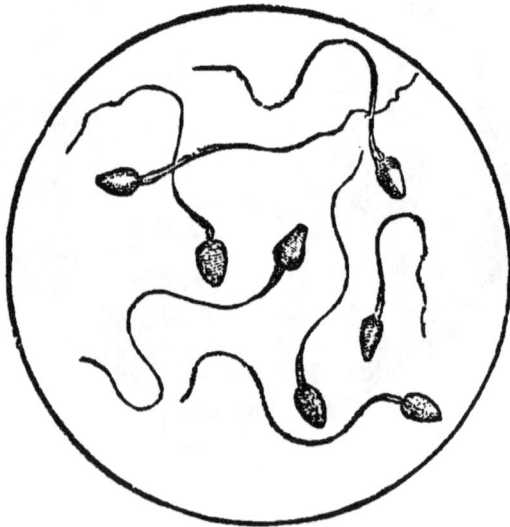

FIG. 5.—Human Spermatozoids; magnified 800 diameters.

very fine granular corpuscles, only $\frac{1}{4000}$ of an inch in diameter, and the spermatozoa, or spermatic filaments, the only and true essential element in producing fecundation. These are minute elongated particles, consisting of a flattened oval extremity or body, and a long slender caudal filament. A small circular spot is seen in the center of the body, and their movements, which are constant and remarkable, being a lashing or undulatory motion of the tail; these animalcules are always present in healthy semen, and were discovered many years ago, but only lately have their true nature and object been fully studied; but with the high power now obtained by the microscope, they have been fully and patiently studied, and all their movements tracked, and they are found to exist in all animals, but each has a peculiar form, and are supposed to be developed from the seminal granules, that finally open and allow the animacules to escape. The size of these minute beings is only $\frac{1}{5000}$ part of an inch, and their numbers are immense, while they are in constant motion, except when in contact with acid or cold, but the movements will return on the application of alkaline fluids or warmth, and while at the temperature of the body will continue for several days in motion, always going straight forward; they do not exist in the seminal fluid until after puberty, and seldom in old age, though in some rare cases they have been found in the semen of very old men. And in all cases where they are absent, *from whatever cause,* the semen can not impregnate, though in every other respect it may be perfect, and the patient in his full health; this has been many times proven, by separating them in animals. The great importance of these facts will be fully appreciated by our readers, when we treat of the subject of impotence or seminal weakness, and that they give me a correct knowledge of their nature and the proper and scientific treatment of these organs when in a state of disease, or disorganization from

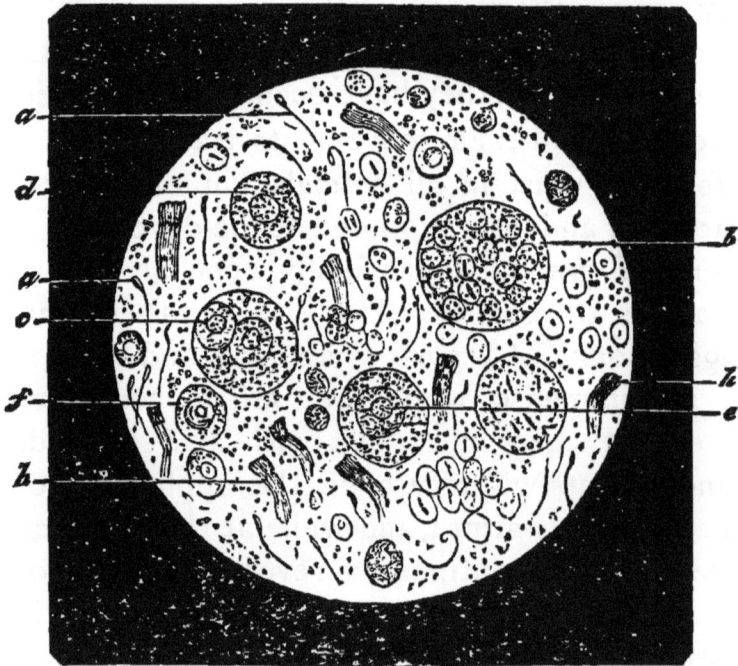

FIG. 6.—Development of the Spermatozoids.—FLINT.

a, *a*, spermatozoids; *b*, spermatic cell containing thirteen nuclei, two of which contain each a head of a spermatozoid developed; *c*, spermatic cell containing two secondary cells, each one provided with a nucleus from which two spermatozoids are to be developed; *d*, *f*, spermatic cells, each with one nucleus; *e*, spermatic cell containing a secondary cell with a nucleus; *h*, bundle of spermatozoids.

self-abuse. The semen thus developed and mixed with the various secretions before mentioned, is found during adult life, and even at an advanced age, and under proper physiological conditions always contains spermatozoids in active movement; but if sexual intercourse be frequently repeated at short intervals, or the parties abuse these organs so necessary to life and health, the ejaculated fluid becomes more and more transparent, homogeneous and scanty, and it may consist of a small amount of secretion from the seminal vesicles, and the glands opening into the urethra, but without spermatozoids, and are consequently without any fecundatory power whatever, as is fully proved by microscopical examination.

It is, in fact, extremely difficult to make a proper examination of them, owing to their being so transparent, and differing so little in density from the fluid in which they are contained, making it not only necessary to have a good and powerful microscope, but to be skillful in its use, when one can readily detect them, though there be very few and very minute; and in the past years I have examined very many specimens brought to me by my patients, and under every variety of circumstances, and also from many different animals. Nothing can well be more interesting or absorbing than these views of minute nature, to " view her stores unfold," and are of intense interest to the physician. The value which is always placed upon this secretion, the work of those glandular organs, the testicles, is rendered very evident by the fact that it is not unfrequent for men to commit suicide from a *supposed* or real imperfection in those parts, and that men who have lost the penis, or who have been castrated from the effect of cancerous or other serious disease affecting the testicles, thereby rendering them forever unfit to enjoy their life from this privation alone, will generally become moping and melancholy, and generally perish by their own hand. How many times do we see the notice in the papers of suicide of young men, that could we trace the secret cause, it would be found in the testicles, or some malformation of the generative organs, either the result of disease, neglect, or self-abuse. Now, in the case of those who have been castrated before they have known or felt the passions that come to us after puberty, they are not subject to the same depression of spirit and wretchedness of mind and body, nor that longing for the unattainable, as those who are rendered impotent after having shared in the happiness and delights of married intercourse. We have an excellent example of this in the eunuchs, though there is a marked difference in their external characteristics, as by this degrading operation, they are more effeminate in personal appearance than those who are in full vigor and enjoyment of manhood. There are also many cases seeking my advice, in which the testicles do not attain their full size and development, and consequently can not secrete the semen perfectly. This is called an *arrest of development*, and intended to show that the organ had ceased to grow before attaining that perfection that they should at the time of puberty. This effect is very frequently caused by an early indulgence in self-pollution. These cases are very frequently seen in a large practice, and there are many cases of grown men whose organs were no larger than those of a boy eight years old. But with proper attention and treatment, with time, I have no doubt such cases can be restored.

Wasting or diminution in the power or size of these organs may occur at any age; the testicle being generally of the proper shape, though diminished in size, and losing its elasticity and firmness, feels soft, and in texture becomes pale, and its blood-vessels appear smaller than when in a natural state; or the organ may undergo what is called fatty degeneration, in which the spermatic cord is affected by extension of the disease, and the nerves and blood-vessels, which are reduced in number. Disease will also affect these organs, sometimes from self-abuse, and then cause them to atrophy, in which it becomes altered in shape, uneven, irregular, elongated, and diminished in size and weight. In the proper glandular structure, the organ seems to be entirely deficient; this is a very serious affection of the testicles, and may come from some of the following causes: as impeded circulation, local inflammation, whether arising from a special cause or transfer of inflammation to the testicles, excess in sexual intercourse, and onanism, are also efficient causes for an atrophied condition of these organs. Also, injuries of the back part of the head have been the cause of atrophy of those organs, and this would tend to support the views of the phrenologists, who contend that the seat of sexual desire is in the cerebellum, which is located at the back part of the head, and between which and the organs of generation there is great sympathy ; and it is an unquestioned fact, that the brain, either in its entire, or in a particular part, does undoubtedly exercise a great influence upon the generative organs, and the desire for sexual intercourse.

CHAPTER II.

ON SEXUAL AND GENERAL DEBILITY — THE EFFECTS OF SELF-ABUSE OR MASTURBATION.

WE open this chapter with true feelings of sympathy for our fellow-beings, and there is no branch of our professional duties to which I have devoted the study of many years and witnessed the heart-rending effects as those resulting from self-pollution. When we know and feel that this pernicious practice was commenced and entered upon at a time of life when one was wholly unaware of the great danger he was subjecting himself to until he had crossed the "Rubicon" and it seemed impossible for him to turn back to the paths of manly vigor. I shall, therefore, feel it my solemn duty to point out these facts in as plain a manner as possible, that all may realize the great and lasting danger and sin they may subject themselves to; to warn them off the rocky coast of Scylla, and at the same time having a care that they are not dashed to pieces and destroyed in the whirlpool of Charybdis—or, in other words, I would show them the sin in glaring colors and the deteriorating effects of that unnatural practice called onanism or

(31)

masturbation, by which persons of either sex, when alone and in secret, defile their bodies whilst, yielding to the unnatural and filthy imaginations, they endeavor to imitate and procure those sensations which can only attend the sexual act. It is such causes as these which combine to destroy, to blight, and to undermine the vigor and manhood of the youth of this great nation, and to whom we may trace with prophetic vision the wide-spread, far-reaching, and desolating plague, which, more than all other causes combined, has peopled the suicide's grave—the asylum for the wrecked and ruined in manhood's dawn.

Any man—be he young or in the zenith of his development and full growth--whose fancy and heated imagination will picture to him in sweet and seductive visions those ideas and desires which delight the excited mind and allures him on to fresher fields of thought and excitement in the unbroken solitude and quiet of his own chamber, and who, in a moment of forgetfulness, yields to the prompting of his passions, has opened the way for many thousands of woes to him in his future years, and will poison his future life and render unhappy and a place of dread, what should be to him and all a place of unalloyed happiness—the bridal bed.

As the mind becoming inflamed with these many thoughts of a lascivious nature, and the excited fancy paints with all the arts and graces of an angel the form and figure of the one adored, and the imagination feeds the flame until every feature, nerve, and ligament is strung to the utmost, and the blood courses through the arteries and veins with redoubled force; the whole frame is aglow with the excitement of the present moment, when the spell is broken, the dream is passed, and all seems dark, only for the waking victim to realize how fearfully he has lowered himself, even in his own esteem and respect, and feels the intense disgust with himself and all around him. Thus it is that the unfortunate victim of this pernicious habit seeks relief or pleasure or perhaps revenge in the solitary haunts, and from which he comes with the determination never to yield again, but only to find all his good resolutions broken and shattered to the ground, and he sinks deeper and deeper in the hands of vice and self-abuse. Nor is there any single vice, disease, or irregularity that in the course of human life ever causes so much sorrow, worry, and anxiety as this. How often do we find it in the experience of practicing physicians that young men have gone on in these practices, little realizing or knowing their great danger, until — united to one adored for a long time and to eternity—they find themselves unable to enjoy the delights of the marriage-bed, as I have seen many such cases, now entered in my books, and they will be noticed in the latter part of this work.

3

And from these facts you may see how so many of our young men of the present day suffer so much from many causes that seem hard to explain, such as pains in the head and back, giddiness, noises and ringing in the ears, loss of flesh, wakeful nights, remorseness and tired feeling on rising, and in the limbs; palpitation of the heart, bashfulness, avoiding company, nervous and apprehensive of some trouble, with loss of energy and will power, a bad memory, and a loss of generative force and power. These are a few of the many symptoms that will come to all sooner or later that will indulge in this pernicious habit; that he has gone so far that he can not produce with these overtaxed organs that healthy cell so necessary for the reproduction of a normal and well-developed being; as he must know how weak and ineffective he must have rendered his generative organs, and then should realize, and may truly ask, "What can I do to be saved?"

This delusive and pernicious habit is first communicated at the many public and private schools, and seminaries, where our youths are sent, and even in many cases at the early age of nine or ten years, from companions or nurses, and before they are at all aware of its dreadful consequences. The absolute importance can not be overlooked that they should have a virtuous education, that they may restrain their unruly passions when they become of that age when they may throw off all parental restraint—as surely it will be, " As the twig is bent, so will the tree incline." So in the period of youth, with the conscience tender, the heart susceptible, the imagination vivid, and the cares of this world, in the far future ; does he receive those impressions, that may lead him on to so much worry and unhappiness, when the mind should be receiving the seeds of useful knowledge, and correct impressions, out of which he will come forth a man with a bright and happy future before him.

Nothing, perhaps, so weakens the intellect as these pernicious habits, which possess the whole mind and body, and prevent their victims from following their daily vocations, rendering them stupid, thoughtful, and dull, and destroying their vivacity, cheerfulness, and health ; bringing on consumption, weakness, and all that horrid train of complaints which make them timid, uncertain, full of whims and ridiculous, and so making it his imperative duty, no matter how seldom the unnatural losses may occur, that he should seek the advice of a competent and skillful physician, that the integrity of his general health, and the power of his masculine capacities, will be both saved from further abuse, and not deteriorated, but restored to their own normal standard ; and unless they are, will explain the reasons why we have so many thousand cases of scrofula, because the *male* spermatozoids are weak, feebly and imperfectly developed, and unhealthy; the result being, that the offspring is born to a fearful legacy, inherited from his parent, who in an evil hour has violated his manhood, and debased his noblest faculties and organs to the enjoyment of a short pleasure of a secret vice, and we also see many, many cases of pulmonary consumption that are due to this pernicious habit, and to this also we may trace the cause and origin of why we should see so many weak, feeble, and rickety children; all due to the uncurbed passions of their ancestors.

And it should be the most solemn duty of all parents, that they should shield their offspring from all dissolute companions, too much freedom of the sexes, and the reading of that highly-colored fiction that forms so much of the literature of our youths of the present day, and the principle of shame and habit of self-denial, fully impressed upon them.

This fearful habit may commit the worst ravages upon youth and manhood, as it strikes at the very root of the propagation and increase of the human race, by debilitating and sapping the springs of life, and that we can help and feel it our duty to warn the youths of the present day from this pernicious habit; and to only seek the paths of rectitude and virtue; though we well know that the temptation and allurement of this mischievous vice, onanism, as it offers such strong inducements for youth, and even manhood to indulge in it, to his future misery and dis-

tress, as it can be practiced in the solitude of the
bedroom, and its effects are so slow and insidious that
we are not aware of our danger until too late; so different for the immediate effects of a night of revelry
and debauch; so for a time this habit is concealed from
all, and this solitary and vicious gratification indulged
in, the evil consequences not known, and so not an-
ticipated, but the undermining process still going on,
until the truth will some day surely assert itself, and
perhaps to the sufferer's everlasting disgrace, and it
does not show itself, in any particular part, more weakened than another, and a failure of memory being
sometimes the first indication of the mischief going
on, and the brain and nervous system becomes weakened and diseased.

The loss of blood, though trivial in quantity, if constantly repeated, is a sure index of a failure of the
vital powers, but the constant and daily loss of this
most elaborate fluid is still more rapidly destructive,
and the general debility produced thereby is very
great; so much so as to affect the brain in its most
important functions. We are taught by physiology, that phosphorus, the chemical element that
enters so largely into the composition of the brain
and nervous system, also forms the essential element
of this vital fluid, and so we can realize the immense
injury that is daily consummated from this repeated
unnatural loss of semen, and shows us the most posi-

tive indications for treatment ; and during a practice
of many years, and in which period I have met every
form and phase of disease, yet I feel bound to say
that in none of the multiform types of professional
derangement, have I witnessed so much of human
agony as in the young man who feels and knows
his unfitness for marriage, when the bride awaits his
coming.

If you have been led astray, do not hesitate to seek
relief and advice without a blush ; do not fear to tell
all to your physician ; it was not your fault, but your
misfortune that you were led onward and onward,
with, at the time, not one thought of wrong that
must lay in your future path; but if you persist in
driving on to destruction, you must take and assume
the responsibility, so that I can advise to all to be
cheerful and full of hope and courage, and strong resolution, to keep from despondency and that terrible
melancholy that breaks the heart and saps the foundation of one's very existence. Seek the true and
proper relief. Let your physician know all your fear
and failing, and in the heart of an honest man
you may be sure of finding relief and comfort, that
in the end, when one's allotted time draws near
to its end, and man's threescore years and ten
have passed away, he may look on the past with
pleasure, and with thankfulness, think of the time
when his feet were led away from the avenues of de-

struction, and are planted on the firm rock of virtue and honesty, that he may look his companions in the face, and say, "I am a man again," and his children shield and guard and protect him, and himself an honor to society and his race.

In pursuing my own investigations into these important and interesting subjects, I have left no means of acquiring information untouched. Besides studying and experimenting as far as was proper in thousands of cases that came under my notice professionally, I have fully experimented on hundreds of animals to the utmost extent humanity would allow. By these means I have ascertained many important facts, and studied the action of many powerful medical agents which could not with propriety and safety have been tried on human beings first.

Functional or sympathetic disability of the reproductive organs generally appears in two forms, *Impotence* and Sterility, which are frequently, but erroneously confounded together. Sterility means a total absence of the reproductive principle, and must always be accompanied by impotence or inability to associate with the other sex, except temporarily, in certain peculiar cases ; but a man may be impotent without being sterile. Absolute sterility is generally incurable, because it arises from destruction or disorganization of the testicles, and it is, therefore, only in the way of preventing the evil, by removing its causes, that we can do good; but impotence *can* be cured as well as prevented.

Besides, impotency is the more frequent affection, and is often merely the forerunner or first stage of sterility, and it becomes therefore the most important subject to consider. It is along this path that the youth travels who would yield himself up to this debasing habit, and so subdued by this wretched infatuation, that while conscious of the change that is taking place, he appears to have lost all power of self-control, or to make a proper effort to recover his position among his fellow-men. Torn by the contending passions of remorse and sensuality, his mind becomes disordered, and himself moody, unhappy, and miserable, and sensibly alive to the impossibility of mixing in the ordinary enjoyments of life, and of deriving from sexual intercourse any of those thrilling delights which are inseparably appended to that act; and he avoids all intercourse with his friends and species, bidding a gloomy farewell to all cheerful society, and the haunts of men, and all the anxieties and hopes of men. The pursuits of business, be it trade, politics, or commerce, are too great for him in his indolent imagination, or too great for his desires.

Imbued with this idea, he will vent his tirade against the world at large and his own feelings, which are only the results of his vices. Thus he becomes isolated and so fearfully changed, the dupe of a lust alike horrible in imagination as well as act, and all his youth and health and the supremacy of his mind, degraded and gone.

Some will continue this habit from a feeling of revenge or despair, as though conscious of its ruinous tendencies; they have endeavored to break the bonds that held them in an iron grip, and to forget this unmanly habit; they have sought the intercourse of women: but unprepared, and their mind in a disordered state, they have found themselves powerless to enjoy the act, and vexed, ashamed, and dispirited, they forego any further attempts, lest they should fail again; and so, abashed, shrink from the gaze of friends, thinking every eye is upon him and reads his innermost thought, when it may be only in his excited and heated imagination; and so continues on in the downward path, until his haggard, pale, and inexpressive face, his dull eyes, and thin, tremulous form, all betray him to the eye of a practiced observer. Too miserable to seek that advice that could restore him to his manhood again, his imagination burns with an unnatural glow; his bodily organs, taxed and weary, refuse to obey the stimulus of that depravity which goads his fancy at night, and in dreams; his broken rest drives him on to another cheerless day. I can not help just here quoting the burning words of Sir Astley Cooper, who, in one of his lectures, said: "If one of these miserable cases could be depicted from the pulpit, as an illustration of the evil effects of a vicious and intemperate course of life, it would, I think, strike the mind with more terror than all the preaching in the world.   The irritable state of the patient leads to the destruction of life, and in this way, annually, great numbers perish. Undoubtedly the list is considerably augmented from maltreatment, and the employment of injudicious remedies."

The trouble is curable in all cases, but to effect such a result, requires a thorough knowledge of the pathology of each individual case, and through adaptation of treatment to meet all the requirements of its unnatural condition, and it is only the physician who has made this and kindred affections a life-long study, whose skill is acquired by treatment of many thousands of cases, that can successfully treat this disease.

It may be as well just here to give the reader some idea of the letters I am constantly receiving in my practice, that they may know how their various symptoms may be expressed, so that they will be intelligent to the physician, and also to show some of the actual symptoms that my patients feel as it seems to them, and also the primary cause of all their trouble. I will not number these cases, but they can all be seen, and many more similar ones in my notes of cases, and on my letter files.

CASE NO. —. E. G. D. writes:

"MY DEAR DOCTOR:—For the past three years I have been suffering from what has been called, by the most prominent man in the medical profession at this place, dyspepsia, with congestion of the liver and other complications, and I have followed his advice, as well as that of some eminent New York physicians, but with no permanent relief, as it has only been to find a relapse in a short time, and yet I have endeavored to carry out all their instructions faithfully, that I might regain my lost health. At that time I had no suspicion of the real cause of all my troubles, until fortune favored me by meeting a confidential friend, who had suffered the same as I, and who had received the benefit of your advice and treatment. It then occurred to me the true nature of my malady, and again I returned to my former advisers, telling them all my suspicions, but only to have my fears laughed at; and though my urine was examined— though how skillfully I can not tell—still I found myself steadily getting worse. Now I acknowledge to you, and to them, that I have steadily masturbated until four years ago, being at present thirty years old, and though I am told that I look healthy, I have constant pimples on my face, and an emptiness in my head, and am troubled with reeling and palpitation if I attempt a round dance, with a constant fear of sudden death, and sudden flashes of heat all over the body. My memory is very defective, though at one time, when at school, very good, and it is very tiresome to me to make any steady application of the mind to reading or writing. Although I sleep well and soundly, when I awake in the morning I do not feel refreshed, but languid and heavy, and would lie in bed all day. I sometimes have emissions in my sleep, and fear that it passes in my urine when at stool. Hoping that, with these symptoms before you, I put my case in your hands, and shall abide by your advice.

"Very truly, etc.,   ——."

On receipt of the above, I at once advised him that an interview was necessary; he called at my office, and when he came, I gave him a most careful examination bodily, and found he was suffering most from nervous prostration, that had deranged all his normal functions; hence, his dyspeptic symptoms, etc. Satisfied that we must look deeper for the primary cause, I examined his urine, and easily discovered with the microscope that he was daily voiding large quantities of semen in the urine; I did not hesitate to tell him the true cause of his disorders, and with confidence promise him a speedy cure, with immediate relief from his most distressing symptoms.

I explained to him the true cause, and he was willing

to abide by my treatment; and I at once put him on a reliable course of medicine, combined with personal treatment that has never failed in these cases. In this case I had good ground to work on, as his body was well nourished and his physical condition good, though who can tell what might have been the fearful results in this case had all his symptoms been allowed to go on, perhaps to mental imbecility; particularly had his circumstances been such as would have compelled him to a life of constant employment for support, as the sedentary habits of a book-keeper or others at like work. The last time I heard from him he was sound and well.

Case No. —. D. O. B.

"Dear Sir:—It is with feelings of great regret and hesitancy that I am compelled to write to you, that you may give me your kind advice and assistance, that I may regain my former health, and hope you will advise such a course which I can adopt without too great a restraint, and that will not interfere with my public duties. While at school some ten years ago, I first learnt the habit of self-abuse. At that time I thought only of the pleasure it gave me, never dreaming that in after-years I must suffer for it. After leaving school I kept up this habit, perhaps only moderately, but at nineteen years of age I found my health very much impaired, and a sufferer from general debility, for which our family physician ordered cod-liver oil and iron ; but I did not tell him at that time what I suspected was the true cause of my trouble. He also advised change of climate ; so I went down to North Carolina, back in the pine-woods of that State, and there I met a very handsome octoroon, but was unable to obtain my desires, and so what I could not have in reality, I sought for in imagination and self-abuse. I have continued this vile habit up to the present time, and though I have tried to break myself of this habit, it is so strong, that even my religious convictions can not overcome my desires.

"I will now try and give you some particulars of myself, and leave the rest for you to judge. I am twenty-six years old, not married, six feet high and very thin, gait stiff, and wants elasticity and firmness, eyes weak and sometimes dark rings below them, and sometimes hot and uncomfortable ; hair dry and thin, and frequently blotches on my face, especially at spring and autumn. I have a good appetite, but suffer from cold feet and limbs, and sometimes sleepless nights ; in fact, all the symptoms of general debility. My memory sometimes fails me, and I am also troubled with blushing ; frequently from no cause. Rather a good liver, but irregular. Like all good things in moderation, and smoke also, and don't believe in sedentary habits if I can help it, so take plenty of outdoor exercise. Now, my penis when flaccid is very small indeed—the end much smaller; and, when an erection takes place, seems inflamed. I occasionally am troubled with involuntary emissions in my dreams, but not very often, but always have fearful erection in the mornings when I awake. My urine is somewhat affected, as it has a strong smell and seems to deposit when standing. I inclose you your consultation fee (five dollars), and hope that you can guarantee me a speedy recovery and restoration to health, that I may rely upon you and will ever be,

"Faithfully yours,   D. O. B."

CASE NO. —.   C. P. V. writes to us:

"DOCTOR:—I should have availed myself of your advice before now; but, like many others, I am suffering for a supply of the needful, but inclose the amount of your fee (five dollars), and should there be any more to pay, I hope you will kindly give me time, as you may be assured I will meet it like a man. My only regret is that I did not know of yourself sooner, that I might have had the benefit of your counsel and advice.

"Now, what shall I tell you my troubles are? Well, I am of medium height, age twenty-nine, pale; occupation, druggist; habits regular; had some nervous debility when in company or if called to mental exercise. I have intense twitching, and my lungs and stomach feel so racked I don't know what to do, until after awhile these symptoms pass off, and I feel easy again, and sometimes drowsy, heavy sensations, which cause my eyes to be dull; cough sometimes, but seldom; and my tongue coated sometimes yellow, but do not think my lungs are affected; but would like to have your opinion. My parents are healthy. Now, when I was quite a boy (only thirteen), I learnt the habit of self-abuse from a schoolfellow, and continued it for three or four years, but I became disgusted with it and gave it up, but find that I am now feeling the consequences of my indiscretion, as I suffer with want of manliness, fretting and foreboding, etc. Am troubled with nocturnal emissions (though irregular), and sometimes a little pain in the right testicle. Void my urine well, and can walk well, and do not mind fatigue; sleep well and eat well, but am troubled with internal morbid sensations. I would like to feel like I felt a year ago. Hoping that I am not imposing upon your time, and that you will relieve me, I shall ever be,

"Yours most truly, ——."

CASE NO. —.

"MY DEAR DOCTOR:—It was a hard struggle to decide to write this letter to you, but having full confidence in your discretion and well-known science, I have determined to put my case in your hands, and abide by the results, as I can not longer stand the agony of mind that I now endure, when thinking of the evil consequences of my own past acts of self-pollution which I have practiced for the past five years; I am now twenty-five years old, five feet ten in height, general health good, my strength is also good; but my appetite fails me very often, by the time I have consumed two or three mouthfuls. I feel a desire to stop, my hunger is quite appeased. Sometimes I feel quite melancholy; at other times I have a super-abundance of spirits. I am not much subject to pain, but sometimes after I have committed the act, I have felt a slight pain in the left testicle, and at other times in the passage of the penis; at other times I have felt a dull pain in the left side, for two or three hours at once. I am subject to frequent and nightly emissions. In voiding urine I have seen seminal fluid run away from me, thin and unelaborated, especially when straining at the last few drops, and the end of the yard is almost constantly met with fluid that escapes from me; I have had much trouble and thought respecting one thing, and I do not know whether such a case ever came under your notice or not. It is this: the end of the yard still retains the foreskin or net, the glands are not yet permanently denuded. Please inform me by return, whether such a thing will be troublesome if I thought of getting married. Doctor, do not be afraid of giving your opinion; let me know the worst; would that you could cure me of this horrid malady. I shall know

4

no bounds to my feelings if you will kindly cure me, as I am suffering from impaired memory, and forget myself altogether, being unconscious of other persons being present, and all to be attributed to this horrid practice. As I think I know others suffering from the same trouble I shall spare no pains to bring them to you for relief, as I expect to be. I inclose your fee, and shall ever remain,

"Yours, etc.,   ——."

Such, my dear readers, are a few of the many letters I so constantly receive, all dwelling on this most debasing sin, and as there are so many different minds and so various are the symptoms, that I must leave many thoughts to be filled by my readers, only asking them to be plain, and distinct in all their letters, that I may make a correct diagnosis, and treat them accordingly.

There is no act that will so soon and so easily become habitual—as when tobacco and spirits are used it requires some time before one can accustom himself to their use; but even in the first essay of self-pollution does one feel a new, wild, and intoxicating delight, to which the very secrecy by which it is committed only adds to its excitement—and so the sufferer in the future, once having given himself up to this pernicious habit, has only a life of misery to look forward to, as it is too late to retrace that step, to blot out from the memory and conscience those new

ideas and sensations that lead one on step by step;
the inward monitor crying against it; but his warnings
not heeded, the conscience becomes seared and hard-
ened, till its feeble voice, from oft-repeated sin, becomes drowned in the mad and urgent calls of unnatural passion, and thus the mind depraved and
losing its governing power, and uncleanness having obtained the mastery over the heart, it pur-
sues its victim with lustful conceptions, at all times
and in all places, and so we pass on, the nervous system disordered, and the brain may lose its balance,
and one passes from disgust to indifference, misanthro-
py, and melancholy, perhaps to madness—and even
the gait of persons who practice this vice, becomes
peculiar, so that one accustomed to this trouble can
tell them at once, and one must not think that his
secret is all his own; but may this alarm him of the
danger he runs, that he may stop his pernicious habit,
and seek the proper medical advice that will relieve
him of all his anxiety and worry of mind. So the
solitary sensualist is the victim of the worst, most
unbridled and tyrannical lust that the imagination
can embody; every fair and virtuous countenance
that is new to him, only inspires him with some
filthy idea, and excites him on, in secrecy, to fresh
excesses, and the nervous system will sink under the
rapid, unnatural whirl and denied repose. He is tor-
tured by his anxious thoughts, and becomes moody

and melancholy, not like the moping, melancholy lover, as his woes are but natural and rational, and can soon find an easy termination, by the fulfillment of all his hopes, in a bright and virtuous union.

Whenever the virile capacity evinces the least degree of decadence, it should excite immediate concern, as the process of decay is in all cases a progressive one. Every case of partial impotence, sooner or later becomes complete, unless arrested by the skillful physician; and no confidence should be placed in the recuperative - powers of unaided nature to effect a restoration of the vital powers, as it will only end in disappointment, and the only true course to pursue is to seek the advice of those who have made this subject a life-long study, and where they will meet with proper sympathy and advice; he may interpose the required treatment to arrest the process of disorganization, and start the process of repair and future health, and to imbue the organ of reproduction with renewed vigor.

We would now impress upon the minds of the reader that having given many of the symptoms that he should look for in his debility, let us see what may be some of the results to follow in the train of symptoms, as the natural circulation of the blood in the brain is disturbed by the baneful habit of self-pollution, and by its drain on the nervous system thereby causing great debility from these too frequent and unnatural exertions, which have a tendency to cause atony, and a palsied and enfeebled state of the male organ, and so rendering it unfit for the natural and pleasant act of virtuous coition, so that frequently, in attempting the act of intercourse, no matter how great the previous excitement, the individual finds himself impotent and powerless, and partly from fear and anxiety, but mostly from absolute weariness and flaccidity of the penis, when, perhaps for many hours previous, it was in a state of constant and painful erection, from the sufferer's previous lascivious thoughts.

All these observations, we may find, were also held by the ancients; as in the time of Hippocrates, who wrote under the title of Tabes Dorsalis on the ills of self-abuse: "This disorder arises from the spinal marrow, and those who are given to unnatural enjoyments are afflicted with it. They have no fever, and though they eat well, they fall away and become consumptive. They feel as if a sting or a stitch descended from the head along the spinal marrow; every time they go to stool or have occasion to make water, they shed a great quantity of the seminal liquor; they are incapable of procreation, and they frequently dream of the act of coition; walking particularly in rugged paths, puts them out of breath and weakens them, occasioning a heaviness in the head and noise in the ears, which are succeeded by a violent fever that terminates their days." And further he says: "This man will not be capable of propagating his species unless the healing art afford him relief." Continuing, he says further, "That when this distemper continues for a length of time, it assumes various appearances in the constitution, and makes other stages under different characters; if not rightly understood it may end in an atrophy or nervous consumption, when the healing art may be unavailing." And we might quote numerous other writers of the olden and the present time to confirm all we have said, and to picture the evils of this self-debasing habit, in far more glowing colors than we have endeavored to portray.

The effects of this habit, masturbation, is frequently the cause of non-development in the male organ, and also its power of erection becomes destroyed. Can we wonder at this, when we note the difference between that act and the natural one? As in the first case, if the seed vessels, the vesicular seminalis, are not sufficiently distended with the fluid that excites erection, it can be forced by unnatural friction to produce a momentary discharge when nature refuses the necessary firmness for coition. In this way the testicles are called upon suddenly and violently to secrete a thin and worthless fluid, and the nerves of the penis are rendered sensible of an agreeable titillation without the natural adjunct, firm erection. Hence when the votary of self-pollution attempts intercourse with the other sex, there is an absence of firmness in that organ to effect penetration and render the act complete, or he may partially succeed, only to have a premature emission, or no sensation of pleasure.

In the above pages, I have tried to give an idea to the willing mind, sad though they may be, of the many consequences of those unnatural sensual enjoyments, so plainly and decidedly the reverse of that transporting emotion enjoyed in the love and caresses of a pure and virtuous union, and which counterbalance, in some measure, the luxurious fatigue consequent upon rational and temperate indulgence, to which the victim of self-abuse can never enjoy or know, and all these enjoyments can only exist in his imagination but to urge him on to further excesses, while the joy and pleasure of a true and virtuous union to the loved one will animate the circulation, restore the strength, support it, and give to marriage all its home-felt sweetness, that owing to the degradation of his soul, he can never realize this pure interchange of affection.

This pernicious habit, so dreadful in its effect, is not alone confined to the male sex, unfortunately. This indulgence has found its way to the pure chamber of young unmarried girls, learned, perhaps, from older friends. One can hardly realize the great responsibility of those whose duty it is to guard over their welfare, and to whom their morals are entrusted, that they should be watched in all their associations, and the books they read, that they may be pure and white as snow. As the same influence that entails early decrepitude on boys, will have its pernicious effects on our girls, that can hardly be believed, except by those whose profession brings them in daily contact with those who have forgotten their own modest nature, and indulge in such practices that may entail on their future much unhappiness and barrenness, and may result in those persistent discharges, too common in the women of the present day, besides many serious ailments that will undermine their health and very existence.

I will close my remarks on general debility, with a few words of advice and consolation to the sufferer thereby, and to offer him that sympathy he longs for: The virile powers can always be restored, but assuredly only by judicious and appropriate treatment. The physician in ordinary practice relies on his aphrodisiacs, or medicines supposed to excite the amative passions, and to give tone to the failing system; but failure ensues, because such remedies are not proper, and may do harm. We must attack the primary cause, and so restore the paralyzed condition of the genital nerves. Each case will present different features, so that the treatment should be ordered accordingly, and as appropriate treatment will infallibly restore manhood in all cases, every sufferer should engage the services of a physician, and they need skillful and scientific treatment in the use of *direct local remedies,* which none can give unless well skilled in their use, and have the necessary apparatus. These cases are beyond the use of internal medication. And, in conclusion, let me impress upon your mind, my readers, the sometimes fatal results of delay, and if you are anxious or troubled with any of these symptoms, leading on to decay, or that you have, in the thought-less hours of youth, when health and strength were at its height, indulged in these unnatural practices, and now feel its warning voices, turn your steps where you may be sure of proper advice and interest; that, in the course of time, you may be restored to man-hood, in all its glorious feelings, and all its strength and independence; that in the end you may find some true and fitting helpmate, that, with the little pledges of love and affection that will spring up in your path, guide and assist your steps to honor and reward in this life and the one to come.

CHAPTER III.

As the semen or genital fluid is secreted by the testicles, it passes along the vas-deferens in its course through the inguinal canal to the base of the bladder, and is there stored up for future use in the vesicale seminales, to be ejaculated during the act of copulation through the seminal duct, passing out of their openings or mouths to the urethra, and there mixed with the prostatic fluid from the prostate gland, and while stored away in the vesicles the watery particles may be absorbed, so rendering it much stronger and concentrated, and so giving rise spontaneously at certain intervals—if the individual is in full health and vigor—to a natural and physiological desire for sexual intercourse.

Such, then, is the order and state when all the parts have their highest physical development; and, as the act of evacuation is forced and unnecessary, just so does it become pernicious, and so bring on its

(58)

own pathological state, with all its train of symptoms, in all those who may be addicted to excess of venery or that indulge in self-pollution. Hence it follows that all those addicted to this pernicious habit may have the power of exciting those organs to a seeming satisfactory result; when, from the many previous evacuations and the consequent overwork that they have stimulated the reproductive organs to, will result in the ejaculation, containing nothing that can stimulate, and produce the fecundatory principle. Nor may the excitement last sufficiently to complete the marital act, or there may be a premature ejaculation of semen, caused only by the irritable vessels pouring out a thin mucous discharge with no vivifying power, and so rendering the act both unsatisfactory to himself and his companion, who is, perhaps, the one above all others his companion for life, and who must look to him (and does expect from him) all the delights and pleasures of the marriage bed.

From this constant source of irritation the vesicales seminales or receptacles are unable to retain this vital fluid brought to them by the testicles, and so will the victim of his own past errors suffer from all the various symptoms that are the result of self-abuse; and, should this fraction of manhood contract the marriage vow and lead to the bridal bed some pure and lovely creature to enjoy that highest of all our enjoyments of this earth, he will only bring on disappointment and sorrow from having reduced himself to a state of helplessness and impotence; or, if he may complete the act, a thin, gleety drain is all he can furnish, with no impregnating power; and should it contain any of the vital principles, they are so puny and diseased that the offspring can only suffer from such an origin. These various results are but a fraction in the troubles that will come to all those who, from past abuse, have brought on the many forms of seminal weakness, and that require so much tact and skill on the part of the medical man, and fortitude and confidence on the part of a patient, to restore his generative organs to a state of health.

This brings us to the subject-matter of this chapter, called spermatorrhœa, or seminal weakness, which is my description of this true disease. I shall divide it into two classifications; first, we have true spermatorrhœa, or loss of the spermatic fluid, without sexual desire or excitement; and false spermatorrhœa (very similar), in which there are nocturnal emissions very frequent, when diurnal emissions take place on any sexual thought, and a urethral discharge of a glairy fluid that attends defecation or evacuation of the bowels, and when erections with discharge follow the slightest irritation—such as that produced by riding or walking, from the friction of trousers, etc. These

cases are very frequent, and, as such, they are the
results of masturbation—from which and to guard
against, I would warn all my readers. These are but
the first symptoms and warning voices that precede
true spermatorrhœa—a result that will surely follow
with all its fearful and debasing train of symptoms;
and so these emissions, though they be attended
with a pleasurable excitement, should be checked
by the proper treatment for such cases, as they will
only lead on to the true disease, with the whole
genital tract in a state of hyperæsthesia, or excessive
sensibility, due to false spermatorrhœa, that will
surely lead on to the true variety.

And what may we expect? First, a stain upon
the linen perhaps, or a strange feeling of weakness,
and perhaps an entire absence of the natural erection,
which generally occurs when first waking in the
morning, and though the sleep may not be broken
to the sufferer, conscious yet, he finds the evidence
of his weakness and disease in the morning, and can
he then wonder at his feelings of lassitude and despondency? and should he not seek at once the proper relief and advice, that he may have the most
prompt and energetic treatment for the relief of his
impending mischief—true spermatorrhœa? These occurring very frequently, the "emissions denote a morbid erethism and weakness of the organs of generation,"
and may occur nightly or several times during the

night, and as they cause their debilitating effect, so may they increase in number, even without an erection, and with little or no venereal excitement—the patient only aware of his weakness from the seminal stains.   This is the result to all those who have been addicted to venereal excesses or the unnatural abuse of the sexual function.   The mind, particularly, if the affection be attributable to unnatural abuse, becomes greatly depressed, and the patient, apprehensive of impotence with all its dreadful symptoms, passes on until his emissions may become diurnal as well as nocturnal, in which the seminal emission may pass backward into the bladder ; it will then show itself in the urine, or may be expelled at the last portion of the act of micturation or passing the urine, and so the patient passes onward, neglecting to seek the proper advice, or staving off the evil day, until he becomes a confirmed hypochondriac, with nervousness, impaired nutrition, lassitude, weakness of the limbs and back, with no capacity for work or study, aversion to society, and loss of memory, with love of solitude, timidity, self-distrust, dizziness, headache, pains, weakness of the eyes, and pimples on the face ; he notices that the testicles are loose and dangling, with coldness of the glans-penis, and short, and perhaps painful erection.   The progress from one stage to another, or from false to true spermatorrhœa, is very gradual, and to the patient almost imperceptible.   It can only be detected by the experienced eye, as the seminal losses are sometimes so hidden from view that we only make a true diagnosis by microscopical examination of the urine.

In the next place, through these practices of onanism or venereal excesses, the testicles acquire a morbid habit or sensitiveness that on the slightest local irritation their secretive powers are called into action, and so by keeping them constantly employed they become impaired, and are no longer fit to supply and expend that most vivifying fluid at proper intervals which makes us truly men, and that we may enjoy during an honorable and natural life. But if the youth, in all his ignorance of the laws of life and health, seeks to force his manhood, and so causes the too frequent and premature expulsion of the semen, thereby forcing the testicles on to their work before their delicate organization has been perfected for their true physiological functions, this must, and naturally will, take on a morbid action, and become in a chronic state of disease, and seminal and general debility must certainly ensue.

We may quote the words of one high in the profession, and who has devoted a life-long work to this special study. In his description of the symptoms, he says: " The involuntary emissions may occur both day and night; they take place as often as three or four times a week, and not unfrequently two or three times in one night, sometimes with and sometimes without dreams, though it is probable that the dreams occur in all cases, but are at times forgotten. On leaving his couch, the patient feels much exhausted, and frequently finds that he has perspired much during the night. A trembling weakness has seized upon his limbs ; he has no appetite for the morning meal, to which a healthy appetite addresses itself with so much good-will. The diurnal emissions happen at urinating and at stool, and in almost all patients we find more or less dribbling away. In some it is perceptible by palpable drops more or less frequent, and in others by a continual moisture of the lips of the meatus urinarius." So we can examine the most palpable and obvious outlines of this wide-spread scourge, consider the mild, yet insidious encroachments of this disease, giving us its first warnings in those changes and symptoms which are the sure indications of a vile and pernicious habit, for which the patient at first is not to blame, but should heed the warning voice and seek the proper relief. Hence the physician, careful not to wound the tender, acutely-sensitive nature, should grant him all that consideration and sympathy compatible with a delicate and refined counselor of the human race.

Spermatorrhœa, although so destructive to both mind and body, is a common affection, that has suffered much from neglect and ill-treatment before coming to the notice of the physician. And why? because the young man afflicted with this malady naturally shrinks from consulting the family physician, and if he do so, it is only for the symptoms, not for the true disease, and so the primary trouble is overlooked or concealed; the patient still continues to suffer. Among the many remote symptoms of this affection, we may mention dyspepsia, in its many forms, arising from gastric disturbance; and also constipation and loss of appetite, that will not yield to the many remedies the pharmacopœia contains at the present day, because this trouble is complicated with or caused by spermatorrhœa, and all the remedies are in vain or worse than useless unless the seminal loss be checked, and its cause thoroughly cured.

These words may also apply to many other diseases to which the human body is liable, no matter what the symptoms, and so making manifest the value of consulting one who has made all the results and symptoms of disease of the reproductive organ his special study and thought, and the patient will have at least the satisfaction of knowing if he is sexually unsound. That he may, if so, be treated accordingly, or if he does not heed this warning he may go on from period to period, taking vast quantities of useless drugs, and all to no purpose. No doubt, each one is functionally affected, and they are only secondary, but are aggravated by being treated as primary troubles, while the

5

real cause is neglected until some organic lesion has occurred.

There is no doubt whatever that many men suffer from false or spurious spermatorrhœa, and from nervous prostration and physical exhaustion, who have never had any true seminal weakness or disorder of the reproductive organs. Hypochondriacs and nervous individuals, who are in the habit of brooding over their ailments, and particularly those who read the exciting and sensational advertisements in the newspapers of the day, imagine they are afflicted with all the symptoms of this malady, especially if at any time they have provoked a cause in their early youth, and come to me firmly believing that they are suffering from some form or other of these diseases, when on careful and correct examination I found them quite free from any disorder of the genital organs, and in such cases there is very great difficulty to convince my patient of the true state of his mental troubles, as his nervous system and feelings had been worked up to a pitch of excitement, and all his symptoms exaggerated by sensational readings.

I have given an account of the symptoms of this malady, and can only advise all those who are afflicted, that in the treatment they must listen to reason and judgment, and exercise their powers of self-control, to seek cheerful company, avoid thinking over their many real or imaginary troubles, and to give up

sedentary employments, and mental occupation, with
change of scene ; and would point out to them the necessity of submitting to a thorough examination, that
the true nature of their malady may be discovered,
and a correct diagnosis be given, and proper treatment entered upon to their future happiness.

As the transmission to the second stage, or true
spermatorrhœa, is very gradual and almost imperceptible, I will try and give my readers some account
of the phenomena observed in this most pernicious
disease. The emissions at first are attended with erection and all the sensation of the erotic state begin in
time to occur, without either erection or sensation,
and ultimately take place during the day when the
bowels are moved at stool, or the urine passed, and
sometimes with the excitement caused by the presence of female society ; in bad cases there is an
almost constant discharge or oozing of the semen and
loss of the power of retention, owing to the lax condition of the mouths of the seminal ducts in the
prostatic sinus at the veramontanum, and the patients are much worried from their continued nervousness, incapacity for business matters, with depression
of spirits, aversion to society, etc., and the loss of
semen that occurred with sensation having stopped,
they do not realize that they have entered upon a
much worse phase of the disease, and that the loss of
semen that was formerly only occasional in nocturnal

emissions, is now almost constant, it being carried off by the urine, and at each evacuation of the bowels. To explain this trouble we find, if the reader will refer to our remarks on anatomy, that as the semen passes from the testes along the seminal ducts or vasa differentia, they have sufficient power, in a healthy state, of retention of the seminal fluid in the ducts and vesicles; and when from the abuse of the sexual functions they become weakened and relaxed and somewhat dilated, so allowing the semen to escape involuntarily on the slightest excitement, passing either outward or backward to the bladder, and so being voided with the urine, may not be noticed; it is only by the marvelous power of the microscope that we have been able to diagnose this most important fact that would have passed undiscovered, and to which hundreds have suffered without relief, or a suspicion as to the actual cause; but now, thanks to the great step science has taken, and to the practiced eye, we can tell the true casue of all these obscure symptoms the victims of spermatorrhœa may suffer from, and that will end in true impotence and perhaps sterility.

That we all desire to perpetuate our species is an established fact. It is one of the promptings of all our most true passions, and is a true function of nature, and as natural as hunger. This desire comes on at a certain age in both males and females, and is called Puberty; a very critical time, at which our frames seem to be perfected, and only need further development to a perfect manhood. At this period there seems to be a total change, both mental and bodily; the boy becomes a man, with his whole appearance changed, and his countenance showing intellect and decision; his voice becomes rough and manly, and his cheeks and lips become shaded a delicate brown; his limbs are firm, his step erect and vigorous, and he no longer delights in those occupations and amusements which he so much enjoyed before.

These changes are all the more marked in the female sex, and the body undergoes still greater alterations as they become fully developed; the bust enlarged, the eyes sparkle with vividness, and the periodical indisposition peculiar to her sex commences, and her girlish playfulness is exchanged for a pleasing bashfulness and modesty, her mind occupied with ideas pure, but strange and absorbing, and she is a woman—" God's fairest work."

Hence one can realize that puberty is a truly critical period of life, and unless a man has passed this time pure and undefiled by self-pollution, he is not justified in entering upon the married state, and particularly if he has any of the slightest symptoms of this trouble, until he has undergone a careful and thorough examination at the hands of an expert before he enters into that most solemn engagement, if he would avoid the most refined cruelty to an innocent yet affectionate woman, to ask his conscience well and truly whether there be any bar or impediment to that union, and if suspicion be even delicately and tremblingly alive, let him *wait* until, reassured of his partially lost powers, he may with confidence meet his blushing bride.

This constant draining of the seminal fluid occurring from excessive venery or from self-abuse, is not always the same in different individuals, as it varies in its effects, some of them being not absolutely impotent, but with their powers very, very much impaired, and so with an undisguised effort they may accomplish the sexual act occasionally, but probably to the disgust of the female; and others, though unprolific, are competent at long intervals, their powers, though weakened, are not altogether destroyed; for with due care, and the really steady employment of judicious measures, this evil may be averted, and the victim in time restored to health and happiness. One should be prompt in seeking his relief, as it is of the first importance that these cases should be seen as soon as possible, and no time thrown away now, by improper treatment, until it is too late for success in skillful treatment, but will leave the blundering, bungling quack far behind. And here we must excite without irritating, and on no account should you tamper or temporize with this infirmity in the idle hope that it will correct itself in time, as sad experience proves the contrary.

We must not pass by genital malformation in its various grades as being one of the causes of impotence, as the prepuce may be adhered to the glands, or so bound down to the frænum that without a trifling operation the act of copulation would be almost incomplete; phimosis and paraphimosis may also render the act inefficient, and perhaps excessively painful to the female, and almost devoid of all pleasurable excitement to the sufferer by these malformations, or they may suffer from some unnatural curvature of the penis, the lateral being the most common, and causing much inconvenience, or the penis may be too small, or in such excessive proportion as to give no satisfaction. We also have the various forms in which the meatus urinarius is not situated at the end of the glands, but placed too far back, and when above or below, called Epispadias or Hypospadias, respectively; these are all congenital, and of course not the fault of the victim, but they will all be treated of in our section of surgical diseases, in which we will explain their various appearance, and point out the advisability of a successful operation at the hands of a skillful surgeon. There are also other causes that will influence the result, among which I may mention the disorders of the urinary and genital organs, as thickening of the bladder, stricture, and the various affections of the testicles and prostate gland, and all cases of so-called hermaphrodism. I might continue to enumerate causes and effects in relation to this subject, that are in almost all cases amenable to treatment, or in which the most positive complaint, by proper ingenuity and surgical interference, be rendered at least very palliative, by which a satisfactory object can be gained; and above all, and in all cases, we would advise an intimate consultation, in which the patient may rely on proper sympathy and advice in his affliction. I am repeatedly consulted by persons complaining of the absence of offspring, due to impotence from some cause, and after careful consultation, have been able to render them very efficient help in this matter, and in many cases enable them to realize their fondest hopes and desires, proving that these cases are not so utterly hopeless as may seem to be the case, and that with proper tact, and the full confidence of the patient, I am enabled to promise them very material relief, in spite of the many unfavorable prognoses of some medical writers on these subjects.

Again, with women I find many causes of barrenness, first that may be congenital, and that in some cases can be relieved by proper treatment; also, we find the most frequent trouble common among that sex—the leucorrhœa, or "whites"—and also the many irregularities of the menses, in which there may be retention or profusion of the menstrual secretion,

with too much frequency in their periods; and all
these troubles we find under the head of chlorosis, or
green - sickness, amenorrhœa, dysmenorrhœa, and
menorrhagia, or excessive flooding; and frequently it
may be due to the peculiar temperament of the females, shown by aversion, reserve, or indifference, and
which renders them unsusceptible of anything more
than a mere passive submission to their husband's
embraces, and so may convert love to hatred, and
make the bridal-bed a couch of thorns instead of one
of roses and pleasure; these things may be amenable
to proper advice and treatment. There are other
causes of barrenness of congenital origin that may affect the womb, or ovaries, or entire closure of the
vagina itself, and that it would be impossible to give
a satisfactory opinion about without a close and perfect examination.

Before proceeding to the treatment of this most
interesting subject, as well to the patient as the physician, I would state that a careful examination of all
these many symptoms, which I have endeavored to
portray as succinctly as possible, show the many in-
terruptions to sexual conquest that may occur, and
to explain in plain and forcible language the results
that a sufferer may look forward to without proper
advice and treatment, and that he may see the true
cause of all his sufferings, and to point out to him the
way of relief; that in the due course of time he can

truly say, " I am a man again " among his fellow-men ; that he should not continue to suffer while all around him seem full of vigor and health to enjoy those precious gifts that man has been endowed with ; and to this happy state it has been my blessed privilege to bring many a poor sufferer, coming to me with his mind racked with melancholy, from the effects of his early, and, at first, innocent, indulgence of a vice contrary to the laws of God and man.

CHAPTER IV.

TREATMENT OF SPERMATORRHŒA, ETC.

IN entering on this subject with my readers, I would tell them that the success by which we would terminate our treatment will depend, in a great measure, upon himself, and must prepare the most determined resolution to follow all the minute directions in their most literal sense; to determine that hereafter he will be master of his own passions, and not allow them to control him in any way, or else he can not enjoy peace or happiness; and when this mastery is gained, then, and not till then, should he come to us for that peculiar treatment so particularly suited to those infirmities. So my object will be at first to restore the organs as far as possible to a state of health by abstinence, and to renew and strengthen the general health. And to accomplish this object, he must first drive away all melancholy, foreboding, and unnecessary doubt, assisted and guided by gentle and judicious advice, that he may look forward to the "bright silver lining" with hope and happiness, and to keep his thoughts busy, if possible, with sober business and worldly affairs, and indulge in a reasonable amount of amusement, the cultivation of cheerful and pleasant society, and to drive away all thoughts that tend to excite the passions and drag him down only deeper in the mire of trouble and dismay; his reading should be light and varied, and somewhat studious, as those of history and some of our best novels, and also the daily attention to the news of the day, in those mighty organs that circulate throughout the world, the daily newspapers; and as he would choose his reading, so should he be even more particular in the choice of his daily companions, not those whose daily thought and conversation only tend toward lewd and indecent subjects that excite the mind and degrade the soul; as it has been truly said, "Show me the company a man keeps and I can tell his character;" as one must be affected more or less by the intimacy of another, be it for better or worse; and I regret that young men of the present day are too apt to pass away valuable time in loose and filthy conversation, showing a foul heart and a degraded mind, not fit for the association of pure and honest-minded men, whose conversation and example would tend to lead one in the paths of virtue, and to the homes of peace and happiness, and in such places may one seek the society of pure and virtuous women, who, by their many graces and pure presence, would lead him to a higher and nobler life.

How many young men with spirits high and a feeling to strike out for themselves, leave the precious influence of their paternal roof tò meet the cold sympathy of " strangers in a strange land." At this critical period how necessary it is that the parents should secure for them a residence in some well-regulated family, that they may continue all their former home influence and comforts, and not to leave them to seek the solitude of hotel life, amid all the temptations that may beset them in their choice of companions and associates. Little can we judge of the many troubles neglect of these simple precautions might cause, and from which many a young man has dated his downfall in the path of debauchery, to end in violent dissipation. And while exercising the mind in these pleasant paths, let us not forget the body in all its perfect form, that we may strengthen and develop one with the other by suitable exercise and change of air and scene, when possible, with walking, riding —but only moderately on horseback—and other outdoor exercises.

There is no disease in all the text-books that is so slow and insidious in its onward progress as this, and so the sufferer passes on from day to day, seldom feeling so ill as to make it necessary to seek proper medical advice, and so he passes on till, by repeated and continued prostration, he has injured the generative organs and perhaps the constitution of the individual, and these symptoms so diversified in their nature, that result from self-pollution, resemble in many respects those diseases of an entirely different nature, and by which the ordinary physician's attention is engrossed, thereby leading him away from the true cause of the patient's complaint, and disappointment is the result of his treatment, so the evil work proceeds until when its true cause is suspected it may be too late to restore the organs to their natural vigor, and then the case demands and should receive our most careful and considerate attention, and then will the patient be disappointed in all his efforts at relief, with the sedative powers of leaden girdles, the cooling properties of nitre, or the anti-spasmodic virtue of camphor, and should turn his step to the proper paths of skillful advice.

So in all our treatment we should first seek to remove the cause, to go back to the first steps from the paths of mankind, and to teach the mind to elevate itself to a higher and nobler sphere, that he may not relapse to his former darkness, but become a convert to chastity and honor—and will the usual dosing of tonics ever be of much service in the treatment of seminal weakness? As the true seminal fluid is obtained principally from the most vital portions of the blood, and is manufactured and elaborated by the peculiar functions of the testicles, and so in an excessive drain on this vital fluid, we find the whole

system weakened and debilitated, which renders the
generative organs inert and useless, and this important deficiency can be supplied only by the employment of such peculiar combinations as will exert a
nutritious, warm, and invigorating effect on the system, and prevent and replace the excessive drain
upon it, so to impart tone and strength to the seminal vessels, and not to overdo and produce irritation;
nor should we over-excite the generative powers, but
only exhibit those remedies that will remove the
proximate cause of all this disease and debility, and
restore to the system its lost energies, by the administration and application of those remedies that
years of practice and experience have proven to me
the only true and correct method of treatment, especially all those cases of both true and false spermatorrhœa, that I have endeavored to explain in these
pages—and just here I would like to say, that I have
made the subject of this disease a matter of deep
and constant study and thought, and that I believe
I have adopted and discovered a *new method* by
which I can act directly on the mouth of the seminal
ducts, at their opening in the prostatic sinus, with the
most powerful tonic known to science, therefore giving
tone and strength to this relaxtion and dilatation of
these mouths or valves, and their efferent ducts of
the vesiculæ seminales; this result is the most necessary to secure, and it requires a most familiar knowl-

edge of the anatomy of the parts and all their intimate relations to the important vessels and nerves
surrounding the urethra in its membranous portion,
and to combine with the scientific local treatment,
the administration of such drugs as may exert a
most powerful influence on the system at large, that
we may increase and multiply the vivifying powers
of the blood, that as it circulates through these important organs, it may nourish and restore them to
their former vigor and strength.

The ordinary modes of treatment are many and
various, but all seem to be about the same, with their
principal dependence on the common aphrodisiacs;
and that in their permanent results may be much
more dangerous than beneficial, as they act like most
strong stimulants, whose action is but temporary,
only to be succeeded by a depression of the vital
powers, and may pass onward to a confirmed and incurable impotency; for this reason they should seldom be employed, unless in a few special cases in
which they may benefit. I can and do give such
medicines as will—first, acting slowly through the
constitution, invigorate and impart strength to the
genital organs, and will be permanent in their effects,
though gentle in their action; so that they can not
possibly harm the system, but strengthen and develop it in all its parts, and it does not make any difference if the sufferer be afflicted with false or true

spermatorrhœa, they are adapted for each condition; when, combined with the proper local treatment, my *new method* for this disease, adapted and perfected by myself, so that with steady perseverance and an honest confidence, I can assure all who will intrust their cases to my judgment before it is too late, an honest and speedy cure and restoration to manly health in a shorter or longer period—according as the peculiar features of the case are presented to me.

And in all cases in which these nocturnal emissions or diurnal pollutions may affect or undermine the general health of the patient, we must first check the constant and injurious drain on the genital system; and, having accomplished this result, I endeavor to repair the mischief done by removing all sources of irritability, and by direct application give the seminal duct and vesicles that power of retention —the loss of which was the sole cause of all the trouble, and by removing which I abate and restore all the normal functions of the mind and body, that they may continue in a state of health, and that their possessor, once the victim of so much bodily and mental torture, may henceforward be a man possessed of all his faculties, and in time the progenitor and support of a healthy and intelligent family.

6

CHAPTER V.

MARRIAGE—ITS DUTIES AND EXCESSES, WITH SOME REMARKS ON THE DISEASES, ETC., OF WOMEN.

IT is a divine law that man should seek woman in marriage, that he may fulfill the decree to " Increase and multiply," and so form a bond of union between the sterner sex and the tender sympathies of woman. These natural desires and promptings of nature cause us to leave homes where we have passed the sunny period of childhood, the gentle influence of loving parents, and all the sweet and chastening influences of the home circle, to seek a fame and fortune perhaps in a strange land, and with us to take that well-chosen one to ever love and cherish, and who should be to us " the one and only one."

This is the great and lasting turning period in a man's life, in which he must not only seek a fortune, but should seek to perpetuate his race with beings endowed with every earthly attainment. Hence, to persons properly constituted—both bodily and mentally—there can be no greater happiness than that derived from mutual intercourse, prompted by the mutual love and endearments of an affectionate

couple " all in all to each other." How different from
the embraces of the frail one, who only offers the ve-
hicle of sensuality for the " filthy lucre " at any time
and place. All her charms and blandishments are
hollow and unreal, and her smiles and caresses the
common property of all who care to seek them. It
is but the excitement of a moment, and then forgotten, and perhaps regretted deeply. But with that one
pure being, who gives herself up wholly to her husband alone, how chaste and pure are her embraces,
and how encouraging her smile; and what man can
be such a cynic as to refuse and pass by such delights!
And to such as would enjoy all these delights, as we
should enjoy them in perfect health and strength,
nature has provided for them bountifully and has put
the enjoyment in his power, will he only stretch forth
his arm and obtain and possess it. But *nature's laws*
will not be pushed aside and trodden upon, nor, at
the same time, we must not be too lavish with all
those blessings she has bestowed upon us. Excessive
indulgence must not drain the cup of pleasure to its
dregs and have it still. Hence, we find a cause of
much unhappiness with those who seek the marriage
bed, and which sow the seeds of so much misery in
after-life.

Not only as it may affect the husband in his sexual powers, but this constant excessive indulgence
has its direct influence on the precious health of

woman, and it is almost the constant cause of all those distressing female complaints so common at the present day and that will so seriously affect their general health and in many cases impair their health beyond redemption, only to linger on until a merciful Providence terminates all their sufferings. So many of these unfortunates, who will travel on their weary path bearing all these burdens, that, from a false sense of modesty, they will not seek the advice and counsel of some able physician. And what are those many diseases, etc., peculiar to the female sex that may be brought on by these indulgences and sometimes the secret habits of girlhood? I will try and mention them according to their most common occurrence.

Fluor albus, or leucorrhœa.—This very common disease is called *the whites*, and is almost universal— occurring in both the married and single, old and young, and may even appear in the infant or the aged. It appears in the form of a discharge from the vagina, and may have the appearance of mucus or pus, or even like green water or milk, the color varying from white, yellow, or greenish; or it may be quite colorless, and the quantity will vary very much in different individuals.

This discharge is very annoying in its constancy, abundance, and irritating properties, and may be accompanied by some constitutional disturbance. In the location of this disease we generally divide it into two varieties—vaginal and uterine—according to the origin of the discharge, and we find it caused by those influences causing congestion, and from preceding inflammation ; of the first or congestive causes, are excessive coition, deranged menstruation, vicious habits, various dissipations of any sort, debility, frequent childbirth, and all those causes which debilitate the general system ; and of the second or inflammatory causes, the various affections of the uterus or womb-disease, as malpositions, and resulting in sterility, vaginitis, vulvitis, and pruritus vulvæ. So in all cases we must first find the exciting cause that produces it, and that must be removed, and we must also prevent this disease from becoming chronic in its nature—as it has a decided tendency to become so, and then the symptoms are much more serious— as the female will complain of pains in the back and a weight in the lower part of the abdomen ; the appetite becomes poor, with palpitation of the heart, and headache, giddiness, pain in the breast, etc. ; the skin feels chilly and the head hot, and hysterical symptoms may become very decided, passing on to continual melancholy.

The treatment of this distressing disease is very varied, as the causes are. First we must attend to the general health, to improve it, and alter the condition of the blood, at the same time using the various

local injections of astringent properties, as infusions
of oak bark, alum, sulphate of zinc, with perfect
rest in all cases.   The best remedy that I have used
in many cases is *galvanism*, by means of a large me-
tallic bougie, and will succeed when all other means
fail.   It seems to impart tone to the membranes, and
effects a change both in the character and the quantity
of the discharge in a very short time.   Leucorrhœa
is, however, in most cases, so complicated with other
affections, either as cause or effect, that we can not
lay down any general plan of treatment, as it must
be varied to suit each particular case.   By pursuing
a certain course with one patient, merely because it
was successful with another, we may make matters
worse instead of effecting a cure.

It should always be borne carefully in mind, that
the mere discharge from the vagina does not always
constitute or indicate leucorrhœa, as it may arise from
other diseases, as an ulcer, abscess, or cancer of the
womb; and hence the great necessity of a positive
diagnosis at the start, that the patient may have the
proper treatment.

The next most common cause of woman's suffering-
I find during my practice, to be the various malpositions of the womb, as it may sink down from its
natural position, fall over forward, backward, or laterally, or it may be bent upon itself; of these malpositions, the falling or sinking downward, called

prolapsus uteri of the womb, are the most common; and the other positions are the different modifications of it, with some different symptoms peculiar to each; but in all these positions it is absolutely necessary to make an examination, that the true position of the womb may be distinctly made out, and then to be carefully and properly replaced in its normal position, and by suitable after-treatment, such as wet bandages and proper hygienic means, until the various ligaments have retracted to their normal length, and so keep the womb in its place in the pelvis. The causes that produce these unnatural positions of the womb are many and various, such as wearing corsets, violent exercise, running upstairs, reaching upwards, constipation, and neglect of the calls of nature, injuries at childbirth, and rising too soon after, etc., and may result in any one of those different positions, which will be shown to the patient by the following symptoms, as pain and dragging sensations in the back, leucorrhœa, constant desire to pass water, constipation, unable to take exercise, stoppage of the menses, or a very irregular flow, and the whole system becoming deranged, and the strength fails, making it so important that the womb should be carefully returned to its proper position in the pelvis, and then to reduce all signs of inflammation by rest and cooling applications, and the retention of the womb in its position by a proper pessary; and I can not urge too strongly upon all my patients the absolute necessity of attending to those malpositions at once, that they may be relieved of so much suffering and misery. In my treatment of these cases I have used with the other means the *galvanic* method, and as I have the most complete appliances, I have been very successful in all my cases, as it seems to have a particular tonic effect on those organs, thereby rendering the uterine ligaments so much stronger to hold the uterus in its place.

Of the many affections of the uterus, we find that the menstrual flow is often very irregular or excessive; but chief among which, I find the affection of *amenorrhœa*, or complete stoppage of the menses, the most frequent. The menses are due to a true hemorrhage, dependent on the process of ovulation, in which once in every twenty-eight days one or more ovules in each ovary will burst its envelope, and entering the Fallopian tubes, by its fimbriated extremity, pass downward to the womb. This process in the ovaries is attended by more or less congestion and nervous exalation, and through the ganglionic system of nerves connecting the uterus with the ovaries, that organ is sympathetically affected and undergoes congestion also, then the uterus becomes heavy, sinks in the pelvis, and its mucous membrane swollen and turgid, with the vessels excessively dilated; and as these minute vessels rup-

ture, we have the menstrual hemorrhage or flow, and
for the proper performance of these normal functions
we must see that the uterus, ovaries, and vagina are
in a perfect state of health and vigor; that the blood
must be in its normal state, and that the nervous
system governing the relations of these two important organs must be unimpaired in tone.

Hence those influences that may disorder any of
these important functions, may check ovulation, the
great moving cause of menstruation, and thus prevent the sympathetic congestion of the uterus necessary to rupture the uterine vessels, or oppose the
discharge of blood from the uterus, and we have
the pathological condition called *amenorrhœa*. This
non-performance of the function of menstruation
may be productive of many constitutional evils, as
chlorosis or green-sickness, phthisis or consumption,
dropsical effusions, etc. Now, the causes that bring
on this condition may be due to some malformation
of the organs of generation, some abnormal state of
the blood, or it may be due to some disarrangement
of the nervous system, some of them having a morbid effect, and others merely opposing mechanical
obstructions. In reference to this subject in which
we have amenorrhœa from atony of the nervous
system, it has been well described by Prof. Hodge,
of Philadelphia, under the name of sedation: It
consists in a decrease of the excitability, vigor, and

activity of the nervous agency which controls the functions of the different organs, and has for its cause physical and moral influences ; some of the functions which are under the control of the ganglionic system are the action of the heart, digestion, peristalsis, and regulation of animal heat. In one leading a natural and healthy life, in the country for example, all these are likely to be normally performed ; but if the same individual remove to a crowded city, lead the life of a student, exhaust his nerve power by late hours, bad air, and mental efforts, all of them rapidly become deranged. He suffers from palpitation of the heart, dyspepsia, coldness of the hands and feet, and constipation. This change usually occurs slowly, but sometimes it does so rapidly, as from a sea voyage, or any very violent mental strain. In a similar manner the processes of ovulation and menstruation are affected by it, in some cases gradually, in others with great rapidity.

There are other physiological causes that may simulate this disease, such as pregnancy, the menopanse or change of life, and tardy menstruation, and should be looked for at first, as in the first we have the various symptoms peculiar to that delicate condition, and in which case it would be very serious to interfere in any way ; in the second, it should only occur between the ages of forty or fifty, though in rare cases it has occurred as early as the twenty-first

year, and as late as sixty or seventy, and may be
diagnosed by the absence of sensations of discomfort
at the periods when the menses should occur. And
in the third or tardy menstruation it must be remembered that it may not come on until seventeen
or eighteen, and mothers should not be worried on
that account, but wait until nature takes her own
time to effect a cure, and commence this normal,
healthy function.

The symptoms and effects of suppression or non-
appearance are numerous and often serious, and may
be either local or general. Among the local symptoms are pains and dragging feelings in the loins and
groins with a sensation of weight in the pelvis and
great weakness in the limbs. Sometimes there is inflammation of the external parts and a peculiar excitement which becomes excessively annoying or
leads to vicious habits. The mind and feelings also
suffer, so that the patient is dull, impatient, irritable,
and melancholy, and so acutely sensitive that she can
not stand the slightest disappointment or contradic-
tion. When the suppression occurs suddenly, the
female often feels many of these symptoms acutely,
and some will suffer instantly from a dragging, bear-
ing-down sensation, or from pain in the back, while
others will be seized with headache and giddiness, or
even faint away; others will be attacked with leucorrhœa, diarrhœa, or inability to pass the urine, and

others again will be taken with chills and fever, and they will sometimes have very *peculiar hemorrhages* from other parts of the body—as the nose, ears, bowels, nipples, etc.　This will occur with the same regularity as the normal flow, and is called vicarious menstruation.

In commencing to treat amenorrhœa the greatest care and circumspection is required.　It may be merely the consequence of some other disease, as in cases of disease of the stomach, heart, spine, and consumption, and also from a congenital condition, in which the closing of the natural passage by an imperforate hymen, or closed vagina, is the cause, and in that case only requires opening, and lastly, if due to pregnancy, we must look carefully for that cause.

In all cases we must make a careful study of the patient's constitution, habits, and mode of life, as in some cases it will only require attention to the general health, keep the mind and body in a good condition, surrounded by cheerful society, and to prevent all morbid melancholy and sentimental dreaminess from reading trashy novels and all undue excitement.

When all such means fail, we must resort to medicines, and they will generally have the desired effect. I have always found the preparations of iron to be the best; but the different circumstances of the case will render different preparations necessary, in which it is best to consult a physician ; warm injections and baths may also be used when indicated. There are stronger remedies known, but are not mentioned here, because they should not be employed in any case unless with the advice of the medical adviser, as they might be used from mistaken notions or for criminal purposes.

All other means failing, I have found *galvanism*, if resorted to in time and in a proper manner, will almost supersede everything else in this disease. I have employed it in many hundred cases of amenorrhœa, and with such uniform success, that I look upon it as nearly certain. In many instances a single application will be sufficient, and in every case, if the simplest means should fail, I would advise galvanism before any stronger means are resorted to or powerful medicines given. The manner of application varies in different cases, as the poles may be applied externally—one to the spine and the other on the abdomen—or one may be applied internally in various ways not necessary to mention here. Neither pain nor serious inconvenience attends its use, nor can any injurious consequences follow, even if it does no good. A person of experience, by daily weighing all the circumstances of the case, will seldom be at a loss what course to advise, by which they may have a return to health and their desires gratified.

The state of mind and feelings also have immense

influence on this most important function of the female, also those cases which arise from the natural
passages being closed. We must first make a positive
diagnosis by examination, and they can generally be
easily and perfectly relieved.

From this and the preceding chapters my readers
will readily see how necessary it is that not only the
husband, but the wife, should be in a perfect state of
health and vigor, that they may bring forth perfect
offspring. And to those who have fallen victims to any
secret disease, and who, fearful that any should know
their troubles, undertake to cure themselves—and
when they have nearly ruined their bodily health,
and perhaps seriously injured that of their "life's companion," they will betake themselves to a proper
medical adviser, perhaps only to be convinced that
had they applied for assistance in time, their consti-
tutions would have been saved from all the ravages
of the many diseases that affect these parts.

No man should dare to enter on the married state
when his organs of generation are incomplete or in-
competent to meet all the requirements of that happy
state. No disorder acts so much and so frequently
against the completion of the marriage tie as the various complications arising from self-abuse during the
years of early manhood, and I can not urge my readers too strongly to attend to these matters before
they have full possession of the mind and body.

Nor should men imagine that because the mind and disposition of a young wife may be modest and gentle, that they would rest quiet under the knowledge of the fact that her husband can not fulfill all his marital engagements, that though they may be as pure and unsullied as the driven snow in the matter of sexuality, still nature will assert herself, and teach them the secret which marriage is supposed to divulge. She will not bear the pangs of constant and recurring disappointments without knowing and reflecting on him who is the cause of it all. I may also turn to woman in all her loveliness, and warn her how necessary and binding it is that she should keep herself free and pure, to be always "the one and only one" to her husband, and to be so, the first and greatest object is her bodily health.

And to both sexes—be there many secret troubles in that inner life so dear to us all, of any nature whatever—I would say, do not put off the day when you will seek and find the proper relief, that you may be speedily restored to health, and confidently rely on the utmost secrecy in my judgment and treatment of all these most distressing affections. And let me caution them not to take the chance of rendering themselves unhappy for life, and their future offspring diseased and perhaps deformed, from the neglect to remedy those many ills that "flesh is heir to" from imprudent habits, and from over-indulgence in that most blessed function with which our bodies have been endowed. They should endeavor at the earliest moment, if they have been unfortunate in their early youth, to seek the proper relief, that they may enjoy all the delights of a perfect life. Nature is absolute in all her laws, and will not be disregarded and disobeyed without inflicting a just and due punishment upon the offender, and which can only be remedied by the strictest attention to all the rules and advice of the medical attendant. So the surgeon should be perfectly familiar with all the faults and excesses of his patients, that he may afford them the most perfect relief and prompt restoration to health.

CHAPTER VI.

GONORRHŒA, WITH ITS RESULTS, SYMPTOMS, AND TREATMENT.

OF the many diseases to which the genital organs are subject to, from man's natural desire to seek the relief of his passions, and by so doing finds relief from an impure connection, we find that *urethritis, gonorrhœa*, or *clap*, is the most common and universal, and though the proper means for such relief may be both beneficial and satisfying, and whereby the marriage bed has so many pleasures and delights, the abuse of this natural function, in the paths of vice and prostitution, can be but unsatisfactory, and sooner or later resulting in some form of these peculiar diseases. It is to the sufferers from this folly that I would point out the causes, symptoms, diagnosis, and treatment, that they may be relieved thereby, and that they may seek such proper relief before their disease has brought on the many unhappy results that will surely follow, in the course of this disease, from neglect or improper treatment. And certain it is that these diseases do exist, whatever may be their origin, and that they are poisonous and contagious, though

7

not epidemical. This poison, generated and transmitted by sexual contact, is of a peculiarly malignant
and destructive nature.

There are many theories in regard to the nature
and susceptibility to the disease of different individuals, as daily experience teaches us; that one man
may be infected and another escape, when having
intercourse with a woman affected with virulent *fluor
albus* or *whites*, and it is idle to indulge in any
theories upon this question, nor do we know why it is,
of all the various mucous membranes of the body, only
that of the urethra, the female genitals, the eye, and
perhaps the rectum, are susceptible of gonorrhœal inflammation, while all others are proof against it; and
though the infecting secretion acts first on the meatus, the inflammation is apt to develop in the fossa
just below, called the fossa navicularis; nor can we tell
why this venereal virus always produces a clap and
not a chancre or chancroid; but it does produce its
specific effects in all cases of impure connection, and is
transmitted from one sex to the other, and so exists
in men and women, and in the former attacks the urethra and in the latter the vagina, urethra, clitoris, etc.

In my classification and description of this disease,
I will first speak of its period of incubation, or
the time it generally appears after exposure to the
infection: this is about from three to eight days,
though there have been cases noted in which it oc-

curred as early as twenty-four hours, or as late as two
or three weeks afterward, but these are very rare ex-
ceptions. Credulous persons will hit upon cases
where the period of incubation has lasted longer still.
Every physician that has attended cases of venereal
diseases, especially with those of the better classes of
society, will have found out that it is much easier for
a patient to confess to excesses perpetrated six or
eight weeks ago, than to those he has been guilty of
recently, and they will antedate their venereal contact ; and so seemingly make the period of incubation
much longer; this is the commencement of a most
painful and troublesome disorder, which is rendered
by frequent occurrence one of the greatest social evils
of the present day.

If we make a section of the urethra during the
period. of active inflammation, we find the mucous
membrane that lines this tube, reddened, injected,
swollen, and coated with a puriform secretion, and at
first only involving the anterior portions of the urethra, thus showing how important it is that these
cases should be seen at the earliest moment, that they
may have skillful and proper treatment. before they
have extended backwards to the membranous and
prostatic portions, and making it so much more
serious in its results, and difficult to cure ; this
inflammation of the mucous membrane may extend
and be accompanied by inflammation and infiltration

of the corpora cavernosa, or may extend and abscesses
or suppuration of the prostate occur; we may also
have the inflammation extending through the lymphatics of the parts to the glands, in the inguinal
regions with swelling, though seldom suppuration.
The most common complications of gonorrhœa are an
inflammation of the epididymis (see remarks on anatomy) and catarrh of the bladder, which will generally
show themselves about the end of the first or second
week, at the time the " par prostatica " becomes involved in the inflammation and by which it extends
to the neck of the bladder, or the mouth of the seminal ducts, and hence to the epididymis.

In its symptoms and course we notice that at first
there is an itching sensation at the orifice of the urethra, sometimes extending over the whole gland, and
a tingling sensation is felt, but so slight as only to
provoke frequent erections and desire for sexual intercourse ; and at the same time the lips of the meatus
are found reddened and swollen, and are usually ag-
glutinated by the dried secretion.   A frequent desire
to urinate sets in and the patients have nocturnal
emissions, and during the day frequent erections,
that may lead them to further excesses.   The secretion increases with thickening and swelling of the mucous membrane, causing a narrowness of the canal,
and perhaps partial retention of the urine, and as the
inflammation extends, this itching gives place to a

burning pain in the urethra, extending from the meatus to the fossa navicularis ; as this pain increases it is
extremely severe, during the act of urination, which
becomes more and more frequent, so that with each
effort only a few drops are voided, and with the utmost suffering; the secretion formerly scanty, tena-
cious, and transparent, gradually becomes more co-
pious, thicker, and purulent, and making yellow stiff
stains on the linen. The lips of the meatus are red
and swollen, and the entire penis also, especially the
glands, seem involved in the process of inflammation,
with tenderness along the entire course of the urethra.

Now the prepuce or foreskin, irritated by the discharge or else owing to propagation of the inflammation, often becomes excoriated and œdematous or
swollen, so that the product of a balanitis or inflammation of the glands is added to the secretion from
the urethra ; and if the outlet of the prepuce is small, a
phimosis is apt to occur, or, should the patient retract
the foreskin and from swelling can not return it, will
then cause paraphimosis. At this stage, erections
occur very frequently, more so than at the start of the
disease, and cause the most intense pain and agony,
from the stretching and expansion that occur in the
organs and that will deprive him of sleep at night, and
to resort to any expedient that he may have relief.
With some the inflammation is not so intense, but
merely a slight running, with some heat and soreness.

This is the case in those of a cold and insusceptible constitution ; others are very severely affected, and as the inflammation runs high and the erections are frequent, it causes *Chordee*, due to the altered length of the organ, and to which the urethra is obliged to accommodate, so bends and curves the organ downward, causing the most intense agony, though this can always be relieved in a short time by cold applications. Should the bladder become affected, the patient suffers from a constant desire to urinate, and can with difficulty restrain it ; at the same time he will have violent pains in the bladder and glands.

There is another complication I frequently see : After the disease has been imprudently treated or neglected, and as the discharge may get much less, and, perhaps, cease altogether, we have the inflammation extending along the vas-deferens to the testicles, and so causing the disease commonly known as *swelled testicles*, or *orchitis*. Patients suppose this to be due, in many cases, to the frequent and incautious use of the injections now in vogue, and is shown, first, by a softness and then pain in the cord, testicles, etc., not very severe, and seeming more like a sense of weight in the testicle. Soon, however, the pain augments, and the epididymis, which is the chief seat of the inflammation, becomes very sensitive to the touch, or the testicle becoming involved, an effusion occurs into the tunica propria testis, so that in a

few days the testicles become as large as a goose-egg
or fist, and even larger, and is now less movable than
before ; due to the thickened condition of the spermatic cord ; and the greater the swelling, so much
greater is the pain, and tenderness upon pressure.
If the constitution be irritable, or if the patient indulge in his usual regimen and exercises during the
first stages of his gonorrhœa, this distressing complaint may be expected. And it is the most painful
and dangerous of the complications of this disease.
So the patient that tampers with himself, expecting
its cure, is exposed to much suffering, and a train of
evil of which he had no conception. Another consequence of this disease, and generally resulting from
the balanitis it sets up, are the common *venereal warts*,
that may cover the mucous membrane of the glans
and prepuce, or invade the urethra itself. They sometimes grow to an immense size, have the appearance
of cancerous penis, and will sometimes perforate the
foreskin, if phimosis be present. The only proper
treatment for them is removal, by excision with the
knife or cautery, and with the proper applications to
the base of these warty growths. These warts are
very contagious, and should be treated very carefully.
We have also a peculiar complication called gonorrhœal rheumatism, coming on at the end of a clap ; it
may attack one or more joints, particularly the knee,

with severe pain and tenderness, and attended with effusion in the joint and constitutional disturbance.

I will now speak of the more remote results of a gonorrhœa or clap, that may show themselves, sometimes after all the inflammatory symptoms have subsided, and the discharge stopped; and these results are *gleet* and *stricture*, two very common diseases, so often met with by the physician who has a large practice in these special diseases.

Generally, after a week or so in the treatment of a clap, if the proper remedies are not used, there remains a stationary, scanty, mucous discharge, which may last for months or years. During the day, if the intervals between the acts of micturation or emptying the bladder, be long, this secretion glues the lips of the urethra together, and when the patient awakens in the morning a tolerably large drop of it has collected, and runs out between the lips of the meatus as soon as they are separated. The stiff stains upon the linen are now of a grayish color, although there is a small, but distinct, yellow spot in the middle. This discharge is called "gleet" or "goutte militaire," and may increase and become purulent if the patient exposes himself again to an exciting cause, and the most frequent are excess in wine, or sexual intercourse, as well as exposure to cold, and over-exertion. In the treatment of this persisting complication, we find it is very frequently due to an organic stricture in some portions of the canal, and will both require the same methods of treatment. I will, therefore, to the readers of this book, that is not intended for medical men, point out the way of relief to the sufferer, that he may see and know a correct opinion of his bodily infirmity, and thereby at once seek the proper means of relief, before it is too late, and all these obstinate complications have set in. A few words of advice in regard to the symptoms and diagnosis of this stage of the disease, *organic stricture*, that may creep on after marriage and produce many serious effects, as the person afflicted in this way is not in a proper condition to effect a productive intercourse with his partner, on account of the impediment to the proper ejaculation of the semen. I have had many consultations on this subject, when, by proper treatment, they have been relieved of their difficulty, and the progenitors of very happy families, from the removal of their incapacity.

We find that organic stricture is "caused by a contraction of inflammatory deposit, situated upon, within, or beneath the mucous membrane of the canal," and this contraction and reduction of the normal passage of the urethra is of a chronic nature, and in nearly all cases can be traced to some chronic inflammation, as a persistent gleet. And this inflammatory deposit will go on so slowly and gradually that the patient is totally unaware of his ap-

proaching trouble and danger, unless he should seek
competent advice, or his attention is drawn to the
fact by a spasm of the urethra, or a sudden attack
of retention of urine, caused by some irregularity or
exposure to cold.

This obstruction may be simply a perforated membranous diaphragm, across the canal, or a narrow
band of inflammatory product, deposited and surrounding the passage, and called *whipcord, ring*, or
annular stricture, and may be general or only partial
from some adhesion of the natural rugæ of the urethra, or some folds of its mucous lining. And these
strictures may increase in their extent, from band-
like to the width of one or two inches, or in some
very severe cases throughout the entire course of the
canal; or the urethra may present several independ-
ent strictures, making the diagnosis particularly difficult. I have seen cases in which there were eight to
twelve, and extending from the meatus to the neck
of the bladder, and in that way result in obliteration
of the urethra, with a urinary fistula, affording a com-
pensatory outlet for the urine.

In regard to the most common locality of stricture,
we are indebted to Sir H. Thompson, for his many
pathological specimens, showing that the usual seat
of stricture, in by far the majority of cases, is situated at the juncture of the spongy and membranous
portions of the urethra, and that it will be found

next about one inch in front of the above, and becoming less frequent as we approach the meatus.

As I have observed, the symptoms of this most
distressing and sometimes obstinate affection, are
very slow and insidious, and, perhaps, not noticed until there has been retention of urine; perhaps due to
some indiscretion in diet, excess in drinking, or exposure to cold; these are the immediate cause of the
retention; and then, perhaps, for the first time, the
patient recollects the fact that other and less marked
symptoms had existed for a long time previously.
Perhaps some chronic gleet has existed for a long
time, or some urethral pain during the act of micturation, and that the stream of urine was divided or
twisted, or the act of urination was prolonged, with,
perhaps, a slight irritability of the bladder. These
many symptoms have failed by themselves to attract
the patient's attention, or make any impression upon
his mind, that would lead him to suspect that he was
subject to organic stricture, and that the retention
was caused, in a diseased urethra, by the spasmodic action of the urethral muscles at the seat of
the stricture, and if it is not discovered by the
sudden retention of urine, it will show itself by the
gradual diminution of the stream of urine, until it
will only pass out drop by drop, causing the bladder
to become very irritable, so that the patient will have
to rise several times during the night, with pain on

micturation, and with tenesmus, and causing prolapsus recti and piles.

Such being the results, in their many forms, of the
simple gonorrhœa, and as the patients differ so in
their constitution and peculiarities, it is impossible to lay down any positive or distinct rules in the
treatment of this disease, particularly as there may
be so many different complications arising, that would
alter the mode of treatment very materially.   I
would, therefore, advise all those who may be afflicted
with this troublesome and dangerous disease, that
they do not delay nor tamper with themselves, by
trying the various nostrums for sale in the shops at
the present day; but that they should abide by the
advice and judgment of the physician, who will give
each particular case his best thought and study, that
he may be perfectly restored, no matter how far the
destructive process may have proceeded, as we have
had patients reduced to a deplorable state, by "lues
venera," from the inexperience or ignorance of practitioners; and our utmost skill has been required to
relieve the ill-effect of his past neglect and unscien-
tific treatment.

There are other matters that must influence the
treatment, and control the final result; for instance,
a man may be engaged to marry in a few days,
when he suddenly discovers that he is suffering

from gonorrhœa ; and he must be cured in a few days, as the marriage can not be postponed ; and in just such cases the skill of the physician is imperatively demanded, nor will the general practitioner try to arrest the gonorrhœal discharge by the ectrotic or immediate treatment, as there is much risk; but in numerous cases it becomes necessary, and if judiciously used, it may be employed with safety, but involves great trouble and attention, and an explicit and exact obedience on the part of the patient ; and from these reasons all cases that present for the ectrotic treatment will be carefully examined, and if there is any risk, I will candidly tell them so, and give them my best advice. Before concluding this chapter on this common and disgusting disease, I can not help warning each unfortunate of the great danger and contagiousness of the matter discharged from the urethra, as, should the smallest portion of this discharge be applied to the eye incautiously, by applying the towel or the finger to the eyes when washing, the slightest particle of that virulent poison will be sufficient of itself to inflame that delicate organ to a serious degree. This is called *gonorrhœal ophthalmia* or *gonorrhœal inflammation* of the eye, and if the proper treatment be neglected for any time, it will result in the complete loss of sight, and, perhaps, destroy the organ, with a liability to communicate with the other eye. As this matter is so important, I must again warn all patients to be very careful in the use of all towels, etc., and to consult a physician at once, should there be the slightest sign of this terrible complication.

CHAPTER VII.

ON SYPHILIS: ITS SYMPTOMS AND RESULTS, WITH THE EVIL EFFECTS OF MERCURIAL PREPARATIONS.

IF I may quote the words of a distinguished author and authority on this disease—he says: "Syphilis is a constitutional disease, the result of a specific animal poison, introduced from without," and in which he has reference to the true syphilitic poison, producing its own peculiar venereal sore; but we find that there are two varieties of this disease, in which we have first, the chancroid, or, the non-infecting sore, and the true chancre; and hence arises the difficulty in all syphilitic affections, to discover whether absorption has taken place, and to what extent the constitution may be infected; as, if the virus has not been absorbed, the disease may be treated with ease and terminate in a speedy and effective cure; but when absorption has occurred, after illicit connection, then there is excessive danger to the parts, with constitutional disturbance; it is for these reasons that I would point out to my many readers, some of whom may at some time have been affected with this virus, the advisability of an early consultation, that they may truly know the extent of their previous infection, though at the present moment all the external signs of the disease may have passed away ; or, if in the first or primary stages, they may know the exact results, and that we must not only remove all the external manifestations, but must examine all the recesses of the system, and root out the poison that may exist in all the "avenues of the blood ;" and unless I can do this, and positively too, we can only afford transient relief, from which the patient may suffer from a speedy relapse, and be liable to contaminate some sweet and lovely being.

How, then, it may be asked, is syphilis to be recognized? Is it to be recognized by its primary inoculation, or is it only to be known by its constitutional effects afterward? It has been already stated by some authorities that there is no form of local sore, chancre, or inoculation that can with certainty be that of syphilis, but we find that in the cartilaginous indurated sore, with enlarged glands, there is every probability of the syphilis showing itself by constitutional manifestations, while in the multiple, suppurating, non-indurated chancroid, we will have no such symptoms.

These two distinctive diseases are most commonly developed on the body, by means of the local sores of a female, producing inoculation, and consequently

similar sores affecting the genital organs, and according to their nature and properties of infecting the
general system, are called chancres, or chancroids,
and we shall therefore endeavor to present these two
forms to our readers, although the distinctive features
are not so distinct that the unpracticed eye would be
able to tell them, and making it necessary to have a
proper diagnosis with treatment in each case—hence
in that of chancroids, or the simple local venereal
sore; we find them the result of venereal contact, and
generally found on the penis of the male, or genitals
of the female, but these may occur on other parts of
the body; in fact, wherever the secretion from a sore
may be applied to another surface. They generally,
when on the foreskin or glands of the penis, have a
red or angry appearance, and spread rapidly, and if
allowed to remain neglected, frequently eat into the
male organ, and may cause its complete destruction.
In this soft, suppurating sore, we find them often
multiple, with an excoriated surface, and neatly
shaped and cut edges, as if it was punched out. It
has an irregular and worm-eaten surface, and with large
amounts of pus or matter coming from it, with a
tendency to spread rapidly; the base of the sore will
be soft, and in nearly all cases is complicated with
a suppurating bubo.

It is uncertain at what time these sores will appear
after infection, as it has been found to vary very

8

much, from two or four hours to two weeks; but the
usual time is about ten days, and the action of the
sore varies with the condition of the patient; as in
some cases of a debilitated constitution they show a
tendency to spread very fast, and in other cases are
very irritable and inflamed.

In our clinical experience of these cases, we find
that the first symptom is an irritation or itching of
the part, and this if it occurs on the glans is succeeded
by an inflamed pimple, small and watery; and which
when bursting displays a rapidly enlarging elevated
sore, hollow in the center and excessively painful and
sensitive, making it very painful to retract the foreskin, particularly when from contact these sores
soon appear on the inside, and when on the frænum,
this part is frequently destroyed. All the mucous
membranes of the body, particularly those of the
mouth and nose, are very liable to be affected, and
may from sympathy extend to the mucous membrane of the urethra, and cause what is called venereal
gonorrhœa. These sores also become phagedenic, or
sloughing, in which they are very destructive, leaving
a deep and unhealthy-looking surface very difficult
to cure.

I will now turn to that most formidable of all the
venereal diseases—the *true chancre*, the virus from
which at once affects the general system; and has
the peculiar power of affecting the unborn fœtus and

the newly-born child in a direct way from the parents. No other blood poisons possess this power, and making it so necessary to have a correct diagnosis. The poison once introduced into the system, manifests its presence in its own peculiar way, by the appearance of a somewhat irregular although characteristic chain of symptoms. They are local and general and run their course, yet do not eliminate the poison, as they may disappear for a time to reappear in some other form, and may lie dormant for years till some weakening influence has depressed the powers of the patient, and thus given rise to some local affection, to which the practiced eye will read with more or less certainty the influence of some passed affection of syphilis. The poison has been stopped, but not killed, and in the weakness of the possessor, reasserts its power.

No other animal poisons appear to have such tenacity of existence. They produce their specific effects in a definite way, and in a regular series of symptoms, and are either eliminated or destroy life; having made their mark and run their course, they cease to act, and are innocuous, their power for harm being exhausted. The poison of syphilis, however, is so subtle that it is tolerably certain most of the secretions of a syphilitic subject are capable of producing the same disease in another.

In regard to the origin of syphilis, we find it is always contracted through inoculation, from a local chancre or sore, the inoculation of some discharge from a syphilitic mucous tubercle, condyloma, or other sore, or even from the secretions of a syphilitic subject; the secretions from any specific sore being capable of producing in another a chancre of any form.

After inoculation a certain time usually elapses be-fore the poison manifests its presence, the period vary-ing from six weeks to three months. In exceptional cases the secondary symptoms of syphilis may appear within the month, or fail to appear for four or five months; but every week that passes after the third without any manifestations lessens the prospect of their appearance, and if they do not appear within six months, there is very little probability of their doing so at all, and the disease has "been nipped in the bud."

The venereal diseases become constitutional by the absorption and transmission of poisonous virus: first perhaps through the primary ulcer to the groin, and afterward spreading itself throughout the entire sys-tem of blood-vessels. The circulating fluid being once contaminated, the various solid structures of the body become gradually affected and poisoned. As . this venereal taint passes through the blood it shows itself by its peculiar symptoms and primarily or first on the skin in the form of an eruption, or upon the mucous membrane of the alimentary canal, as indicated by sore throat, with some amount of fever and constitutional disturbance that usually precedes their appearance; this skin eruption may be only a rose-rash, *roseala*, giving rise to a mottling of the skin or a more lasting staining, and may assume the papular form, *lichen*, or the various primary forms by which skin diseases usually show themselves. These eruptions may last for a few days and disappear or leave a dusky, coppery stain behind of some durability. All these eruptions have a copper tint, more particularly after they have faded from their first appearance, no matter what peculiar form of skin eruption they may assume at first.

As the outside skin is attacked by eruptions (simple and ulcerative) in syphilitic subjects, so the inside skin or mucous membranes are equally involved. " Every form of syphilitic affection of the skin," writes Lee, " has its counterpart in the mucous membrane; but the appearances will be modified by the comparative thinness of the structure, by the absence of cuticle, and by the little disposition that these parts have to take on adhesive inflammation." The mucous tubercle is the most common form as found on the organs of generation, tongue, mouth, lips, nose, throat, rectum, and anus, and occasionally in other parts of the alimentary canal; and these tubercles may break down and ulcerate, leaving deep excavated sores.

Bubos in the groin constitute one of the first symptoms, from extension of the inflammation, though they may be absent; then follow pains in the head, the joints of the shoulders, arms, and ankles, and these symptoms may gradually increase, especially the pain, which becomes so intense that the patient is unable to lie in his bed. Nodes arise on the skull, shin-bones, and bones of the arms, which, being attended with constant pain and inflammation, at length grow carious and putrid. This form of the disease is called syphilitic periostitis, and is due to a gummy effusion beneath the periosteum or membrane covering the bones, and eventually attacks the bones beneath, and in which the pain is of a constant aching character, and is always aggravated at night; and, in those cases in which the bones of the skull may be affected, it may extend to the dura-mater and brain. " The internal organs may be affected equally with the external—not only the cranium, but the brain within it or the nerves; not only the muscles of the limbs and tongue, but the heart; not only the pharynx, but the æsophagus; not only the larynx, but the trachea, bronchi, and lungs, also the liver, spleen, and other viscera."

Malignant ulcers now seize different parts of the body, but generally begin with the throat, and thence gradually extend to the palate and the cartilage of the nose, which they may destroy; and the nose, being destitute of its natural support, falls in, and presents a flattened appearance externally. The hair and scalp also become affected, and the hair falls off from the head and also other parts of the body where it may grow. The nails become unequal, thick, wrinkled, and rough; and ulcers may arise and cause them to fall off.

The inside of the mouth, throat, and nose becomes painful, hot, and inflamed with ulceration; and pustules appear in the roof of the mouth, which become round, malignant ulcers that may rot the bone as far as the nostrils; and this disease, when it exists in the throat, presents a white, slimy-looking ulceration, with a most offensive discharge and fetid breath; the soft palate may be completely ulcerated away or detached in portions, and the upper and back parts of the throat present one vast ulcerating cavity covered with adhesive matter, and the voice becomes affected—sounding hoarse, thick, and low; while swallowing becomes very painful and difficult.

As we progress in the symptoms and appearance of this loathsome disease, we find that it leaves the face very much disfigured, the cavity of the nostrils exposed from the throat, the natural prominence of the face destroyed, and a disgusting ulceration takes their place. From these ravages the disease may extend to the bones, attacking first their lining membrane with hard or soft *exostoses*, and sometimes with or without pain ; and, when it attacks the bone, we have caries, which increasing, the bones become brittle and break upon the least effort, or their constituent parts may be so dissolved that they will bend like wax.

The long round bones, as those of the legs, are generally the first that suffer ; hence, those enlargements on the shins—the well-known venereal nodes —which are in reality inflammatory enlargements and thickening of the periosteum, which covers them, with deposits of gummy material, and may pass on to bony-disorganization.

As these symptoms come on, perhaps a long time after the chancre has healed and all other symptoms disappear, the patient complains in the evening of each day of increasing pains and aching in the legs or in some particular place on them. There is not much swelling at first, and the pain and swelling generally disappear toward morning. Great sensibility of pain occurs at evening again, and the sleep and rest are broken from the recurring irritation and fever.

So does this disease attack the many organs of the human frame, and we also find that it will affect the organs of sense also ; as it will frequently in its after effects show itself by affecting the eyes and ears, and in which we find them affected with a variety of symptoms. As in the eyes, they are affected with pains (*orbital*), with redness and continued itching externally; and internally we have the various forms of inflammation of the *tunics* of the eye, in which the sight becomes cloudy and is eventually destroyed, and perhaps suppuration may set in with destruction of the ball of the eye. The most common forms of syphilitic disease of the ears are those of affections of the auditory nerve that conveys the sound to the brain through that delicate "organ of Corti," and we may have ulceration and caries of the internal and middle ear, with all its train of symptoms, as singing noise, roaring, deafness, and pain.

Having so far given my readers some idea of this destructive disease, with all its various train of symptoms, cause, and effects, that I may point out to them the necessity of prompt and immediate consultation, for the success and rapidity of effecting a cure in all cases depends on my seeing the cases at their early start that they may be stopped at once. At the same time I would wish to give them an intelligent account of the various modes of treatment, and also a few words about the use of mercurial preparations.

There is no remedy in the pharmacopœia that can be looked upon as a specific for syphilis, although there are many that will have a very beneficial influence in causing the disappearance of the symptoms; when correctly combined they will exert great influence in eradicating this disease, and when the system is brought fully under their powerful influence, this poison, so subtle and penetrating in its nature, must give way to the patient's perfect restoration to health. Now the treatment varies very much both internally and externally, according to the peculiar stage to which the disease has advanced and according to the type or peculiarity of the ulcers. As I wrote in the opening of this chapter, we have the two distinct and pathological varieties, both varying so very much in all their details and treatment and in both forms, the great and only important point to be kept in view is to abridge the duration of this disease, so that it will not make any serious inroads on the constitution beyond the repair of the skillful physician.

The first steps and the simplest mode of treatment is the extirpation and cure of the primary sore or chancre, no matter what its variety; and this may be easily and readily effected by either the caustic or incision—that is, cutting out the diseased parts—the former being a very safe method and not very painful, and one I highly recommend and practice; but at the same time it may be impracticable from the surrounding parts becoming contaminated in consequence of the difficulty of entirely removing the diseased places, making it so necessary that we should combine with the local treatment such measures for

cure in the earliest stages of the disease, that we may
prevent the contamination of the system, and pre-
serve the constitution from this most pernicious
disease.

In olden times and up to the present day the chief
reliance of the physician in the treatment of this disease was mercury, and it was given in all its many
chemical varieties; but it is truly a two-edged sword,
by which, although curing the disease, it may leave the
patient in a much worse condition than before; for we
are told by the advocates of the use of this medicine
that "should the salivation be attended with a cardi-
algia or violent pains and torture at the stomach, per-
petual and incessant retchings, and cold sweats, there
is great danger to be apprehended."

Many forms of venereal sores are rendered irritable
and evidently disposed to slough and mortify under
the action of mercury, and there are many of the
older practitioners who can recollect the period when
mercury in the ordinary doses failed to act as a remedy, that it was the practice to increase the dose,
supposing that a more complete course of salivation
with saturation of the system could alone arrest the
rapid course of this disease, due, perhaps, to the unskillful use of these preparations.

That its improper and incautious administration
has been productive of horrible consequences can not
for a moment be doubted, nor is there any subject in

the entire range of medical and surgical science that demands and requires a greater amount of practical skill and determination than to know to what extent, and just when, and under what peculiar circumstances this powerful medicine should be used for the relief of this most distressing and dangerous disease, as we find that so many, perhaps young and handsome, have become self-sacrifices to their own inexperience and incautious use of this mineral drug, which in the course of their disease has shown itself in unseemly blotches on their bodies and unpleasant eruptions of the face.

One of the worst features of this most distressing and insidious disease, and its most important feature, is the almost certain liability of its transmission of the syphilitic poison from the parent to the child, and by which we find these little innocent beings with no fault of their own, ushered into this world with their systems contaminated and their bodies afflicted with some loathsome eruption inherited from their parents that they must carry with them through life. Truly the "sins of the father shall be visited upon the children of the third and fourth generations of men." Infants may be affected in many different ways; the disease may originate in the fœtus or before birth, in consequence of the impurity of one or both parents, and if from the father's contamination, it will affect the mother also.

Infection may happen when neither of the parents has at the time any venereal swelling or ulceration and perhaps many years after a cure has apparently been effected—we can not tell why, but these are well-attested facts shown by clinical experience. Hence we have premature labor coming all too soon—by an abortion or miscarriage, the offspring presenting a puny, feeble, emaciated, and wrinkled form—the eyes red and inflamed, the cry shrill, husky, and wailing; mattery discharges are emitted from the eyelids, copper-colored blotches disfigure the skin of the genitals and hips; the nostrils are clogged, the nails come off—and, in fact, these little ones come into this world utterly unable to bear the trials of life and childhood, and soon sink out of sight in the silent grave—innocent and sinless victims to man's licentiousness.

In thus dwelling on the evil consequences of sensual indulgence and of the ailments of our depraved habits, let me not be accused of wishing to administer in the least degree to the morbid feelings and the curiosity of the idle; but I would endeavor, as far as this book and my abilities will allow, to show to all my readers and to those who may be in any way sufferers, that I offer them a guide to manly health, and to show them the many causes that may pervert it, as well as to warn off many from the rocky and tempestuous coast of self-indulgence and abuse. For these reasons I have tried to point out to man what he may be and what he may look for will he only follow all the laws of life and health, and, on the other hand, to point out to the victim and the fallen one the terrible and truthful consequences of his sin sure to follow if he continues on in the path of self-indulgence, and to point out to him the beacon-light of safety, by which, with timely consultation, he may meet with honest sympathy, and I can promise a perfect restoration to health. As the physical grandeur of man is in a state of perfection when he possesses every bodily and mental function in its perfect and original strength. Every function is capable of increasing our constitutional happiness when it fulfills its legitimate design, and is natural when exercised to a certain extent; but when we pass beyond the bounds of physiological indulgence and trample the laws of nature underfoot, then it becomes unnatural, and consequently disease sets in with its train of various symptoms. So it is with the gratification of our appetites in eating and drinking, and so must it be with the genital system when overtaxed in its most important and elaborate function. All indulgences are therefore hurtful, more or less, in proportion as they are pursued or restrained.

I have shown the dreadful effects of debasing this physical grandeur by the indulgence of self-pollution, and the effects of excessive indulgence in sexual intercourse, both promiscuous and matrimonial—that the health will be seriously impaired, the body exhausted by this wasting cause, and that we must retrace our steps, seek repair and health, that our progeny may be well worthy of him who gives them life and being.

CHAPTER VIII.

SURGICAL DISEASES OF THE GENERATIVE ORGANS.

I SHALL include in this chapter all the malformations of these organs, be they congenital or the result of some disease of the parts, and shall point out to my readers the distinctive features by which they may be known, that they may be encouraged to seek proper and skillful relief, as there are so many unfortunates that spend their lives suffering from day to day with constant worry and agony of mind from perhaps some slight deformity that could be almost instantly and painlessly relieved. How many men are there now, willing and anxious to enter that happy state of wedded bliss, and yet dare not ask the all-important question from the fear that this secret trouble to them will prevent the consummation of the marital act ; so far they are correct, as these malformations do seriously interfere with the perfect and complete act of copulation by preventing the proper ejaculation of the semen. So they will pass their lives, when I feel sure a few words of honest advice in pointing out to them the source and nature of their trouble and explaining to them that they may rely upon skillful and intelligent treatment, and that their parts may be restored, so that they may enjoy and complete the marital act.

Phimosis.—We find this the most frequent of all the genital malformations, and one that is most commonly congenital; or, in plainer words, this malformation existed at the time of birth, or it may be the result of disease, as caused by the excoriations, and œdema or swelling of the parts from the acrid and purulent discharges with which they are bathed in violent cases of acute gonorrhœa, and so render the cure of the primary trouble much more difficult and complicated. Again, we see cases of phimosis that are caused from the irritation and inflammatory infiltration of a chancre or chancroid existing beneath the prepuce or foreskin, thereby preventing the physician from making the proper and thorough application; nor can he see the progress of the primary disease; consequently, phimosis exists as an elongation and swelling of the foreskin with corresponding narrowness of the opening, thereby preventing its retraction over the corona or crown of the glans-penis, and so concealing any virulent and contagious disease that may exist underneath, as we find in the sloughing and fringing preputial chancres.

In the treatment of this malformation it will depend in a great measure upon the extent of the elongation and swelling of the prepuce and the state

9

of the parts underneath, as there are many cases that may be brought to a successful termination by simple palliative means combined with proper hygienic treatment, or they may require some operative interference that can only be attended to at the hands of a skillful surgeon, and so showing the necessity of a consultation, in which the patient may rely on the most conscientious advice and treatment, and feel sure of the relief of his deformity.

We will next take that condition of the prepuce, which is the exact opposite to the one just explained, and which is called *paraphimosis*. This malformation in its causes is very similar to phimosis, but in which the patient is suffering from a very tight foreskin, and, having been retracted over the corona of the glans, can not be replaced, and so by its tense and contracted condition, together with the accompanying swelling of the glands and the parts beneath, causes a strangulation of the glands and the mucous lining of the parts, and, if not relieved, will cause ulceration at the line of strangulation, and may extend to sloughing of the parts, particularly if combined with some venereal disease; and, in the treatment, we must endeavor to reduce this abnormal condition first by simple palliative means, and, this failing, use proper operative interference.

We can not impress too fully on the mind of the patient the advisability of at once seeking proper medical aid, that these matters may be relieved and not pass on until they have produced almost irredeemable damage, thereby rendering the patient miserable and unhappy for the rest of his life.

I would also call attention to the complications spoken of due to malposition of the orifice of the urethra, called *epispadias*, and *hypospadias*, in the former of which the opening of the urethra exists on the dorsum or back of the penis ; and in the latter, or *hypospadias*, the opening of the urethra is situated underneath, and may be as far back as the scrotum, and by which the patient will readily see that the proper accomplishment of the marital act is simply impossible, and can only result in dissatisfaction to both the parties. Associated with these two malformations we may have loss of the penis from accidents, operations for disease, or congenital causes ; and I have seen instances in which it was not more than half an inch in length, and in all such cases there can be no connection, though they may be fathers, as the semen will impregnate if placed within the lips of the vagina. In these cases there is very little to be done by the surgeon ; nor should any man with such a malformation ever marry without the full consent of the female, she knowing the deformity and its possible consequences.

There sometimes occurs a wrong direction of the penis, in which it is bent laterally or downward when in a state of erection. The correction of this will depend altogether upon its cause; as, if it depends upon contraction of the skin from any cause on that side, it can be relieved by a very simple operation; and, when from tumors or swelling of the veins, they can be relieved generally by some mechanical application.

I had a case in which there was a large tumor on one side of the penis due to rupture of some of the veins on that side from long and continued erections during a state of intoxication. It was very materially relieved by cooling applications and a perfect-fitting mechanical appliance. In some cases the penis will be bent downward from the frænum or cord that attaches the prepuce underneath, and will be so short or contracted as to draw down the glands, and so will prevent a perfect ejaculation of the semen, and even prevent connection by causing too much bending or severe pain. In these cases it is only necessary to sever the cords and keep the parts asunder.

In regard to the size of the penis—a subject I am often consulted about, as so many men think their organs too small—it is hard to "draw a line" as to the proper size of this organ, as it varies so in different people. In some it never grows from childhood, though all the other organs may be fully developed; as in the case of a man whose penis was only two inches long, but his testicles were fully developed and his powers and ejaculation perfect. In those cases in which the penis is not fully developed, and the organ (though small) is capable of perfect erection, both connection and impregnation may be effected; still, some improvement may be desired, or more frequently erection either does not take place at all or so imperfectly that coition is impossible. The flow of semen is so imperfect and irregular, that impregnation can seldom be effected, even artificially.

Under such circumstances it is of the greatest importance to produce an increased development, so that both those functions may be performed. It may be a great pleasure to many unfortunates in this particular to know that there are means, even under most unfavorable disadvantages, by which this malformation may be relieved and the organ developed to a proper size; and such is the case particularly when this trouble is of congenital origin; and to those also in which the organ has decreased from some disease, we can hold out to them the hand of hope, feeling sure we can offer them some relief.

This has been a deep subject of study with me for many years past, and I have dwelt on this subject as I believe I have the means by which so many can be relieved. I have treated many cases with very good results—in some cases only by letter—as I can send the proper appliances, with full directions, by express. I have received some very flattering letters from those who have been relieved by the use of these mechanical and manual appliances—the effects of which, under right direction, are of an unexpected and pleasing character. To understand their value and mode of application, one must remember all the anatomical relations and structures of the parts, as it will be remembered that in the phenomena of erection the blood flows in and fills all those minute interspaces in the corporas cavernosa and spongiosum from the arterial supply; and, should there be any fault in their construction or avenues of supply, the erection can not take place. We may have the same result from long-continued excesses in youth, in which the arteries seem to lose their power of forcing the blood within the cells, and so, from want of being filled, decrease in size and close up more or less, causing the organ to shrink. We may have the same result from continued suppression of the sexual act from non-use.

Therefore we must endeavor in all our treatment, to open and develop these cells, causing the blood to flow into them freely, and so increase their size, and dispose them to fill spontaneously from natural excitement. It is necessary in all cases to use some stimulating lotion, as well as shampooing and rubbing the parts freely to assist the sluggish circulation, and attract the blood to the parts; at the same time the skillful application of the *Congester* will develop the organ to a size that the copulative act can be complete. This congester, though it is not used generally in self-treatment, I would advise all to be very careful in its use, for without proper directions and supervision, it may do great mischief, especially to those who are too anxious to force nature on in her delicate work; hence we may have a rupture of the cells on the penis itself, but the patient with proper guidance may do the shampooing and rubbing himself, though not as perfect as they should be done; but in the application of the congester I would advise all to have that skillfully applied, as in some cases it is necessary to have them made to fit perfectly, or they will fail. In my notes of cases I will endeavor to illustrate several cases that have been under my care with the very flattering results I have gained thereby.

In speaking of the diseases of gonorrhœa, or clap, and of syphilis, we mentioned a complication arising in these diseases, called *adenopathy* or *bubo*, due to an extension of the inflammation to the inguinal glands in the groin; these glands are a part of the lymphatic system, and serve to convey with the lymphatics, the lymph that is collected from the different parts of the body to the venous system, that it may be again elaborated, and the effete matter

thrown out. Now, from the intimate relation between these glands and the various parts of the penis,
as that organ is very richly supplied with lymphatics,
the inflammation readily extends in the course of these
vessels, and the result and amount of inflammation will
depend on the cause, as should the infecting material
not be virulent nor constitutional, as is the case of
gonorrhœa, we would probably have sympathetic
bubo, occurring generally in the upper part of the
groin; these are readily amenable to treatment, provided they are not neglected, in which case they may
suppurate or matter form, thus making them very difficult to cure, so that I would advise simple fomenta-
tion and then to put the case in the hands of a
physician.

Hydrocele, or a collection of fluid in close connection with the testicles or spermatic cord, in most
cases is due to some inflammatory affection of the
serous membranes covering these important organs,
and in which we have thickening of the tunica vaginalis, with deposition of a serous fluid in these cen-
ters, but sometimes due to congenital causes; this is
albuminous in its character, as all other serous effusions, and in inflammatory action is very much increased. In the symptoms of this common affection,
we find that it appears as a painless swelling, and as
an apparent enlargement of the testicle, with very
slow and unequal growth and variable size, with a

smooth and uniform surface, and more or less tense
and fluctuating feeling ; it is always movable in the
scrotum, and as a rule can be demonstrated to be
distinct from any abdominal connection ; with the
testes clearly made out, at its posterior and upper
portions, by the testicular pain on pressure, or by the
absence of translucency at one spot, as the tumor
will, as a rule, transmit light when its coverings have
been stretched. In the treatment of this surgical
affection, if of the congenital variety, is very simple,
as only some simple lotion is needed, as the hydro-
chlorate of ammonia with tonic medicine, by which
we may expect an entire absorption of the encysted
fluid ; acupuncture has also been recommended, but it
will be necessary in nearly all cases, particularly in
those of a chronic nature, that *tapping* should be at
once proceeded with, in the hands of a skillful surgeon. This will always be found a very successful palliative measure, as in course of time the sac will again
fill up, and should the patient desire it, we may pro-
ceed to the *radical cure*, by which the sac, after tapping, is injected with some irritating fluid, and hence
setting up a process of adhesive inflammation, by
which the sac is entirely closed, and thereby preventing the accumulation of any more of this serous fluid.

There are many other diseases and complications
of the genito-urinary organs that will require surgical
interference, such as those of the *kidneys* and the

bladder, but that in their course, symptoms, and treatment are too complicated for self-diagnosis, or to be explained in a work of this kind ; and I would therefore advise, that in all cases of suspected trouble of any of these important and useful organs in the human economy, that they should at once seek some proper and skillful advice. That I have treated many such cases, my note-book will well attest, and with an unvarying success that has been both flattering to myself and to the lasting and permanent relief of my patients, by which they have been again able to enjoy all the pleasure and delights that these organs are so well fitted for.

As all these troubles of the genito-urinary tract do have their most positive influence on the mind and health, thus rendering the victim morose and unhappy, and seriously interfering with the proper secretion and elaboration of the seminal fluid, as of all the organic functions we find that of secretion is one of the most strongly and frequently influenced by the mind. The secretion of tears, of bile, of milk, of saliva, may all be powerfully excited by mental stimuli, or lessened by promoting antagonistic secretions. This influence is felt in full force by the generative organs, " which," writes a distinguished author, " are strongly influenced by the condition of the mind. When it is frequently and strongly directed toward the object of passion, these secretions are increased in amount to a degree which may cause them to be a very injurious drain on the powers of the system. On the other hand, the active employment of the mental and bodily powers on other objects has a tendency to render less active, or even check altogether, the processes by which they are elaborated."

It will thus be seen that as all these various malformations, be they congenital or the results of some neglected or ill-treated disease, must have their influence on the mind from the thoughts constantly dwelling upon these infirmities, and through the mind have an almost direct action upon the secretion of the seminal fluid ; and as this function remains in a perfect state of health, so will the nervous system continue to influence all the important functions of the body.

In my treatment and care of all these peculiar diseases, I have made it a rule to do no more than is absolutely necessary for the speedy relief and prompt and safe cure of their existing trouble ; and they may feel assured that they may put their case in my hands with the full confidence and assurance of the most speedy and scientific relief possible, and that all my necessary operations will be performed as painlessly as possible, as I have several of my own inventions that will render them so, and by which I can do these various operations with almost no inconvenience to all persons who entrust their cases to my care.

CHAPTER IX.

SELF-DIAGNOSIS; OR, HOW A PATIENT CAN TELL HIS SYMPTOMS.

As there are many of my patients and readers of this book who will ask the question, How can I tell what disease I am suffering from? and how shall I write or give the doctor an intelligent account of my sufferings that he may make a proper diagnosis and treat my case properly? I will try and show them how, in as simple a form as possible, by arranging the many symptoms in a tabular form, that can be almost answered by a simple Yes or No. Now, there is a great variety in the symptoms, and they are very numerous both in number, in nature, and in degree, and I shall therefore divide them into those affecting the generative organs, first; those affecting and influencing the muscular, circulative, nutritive, and respiratory systems, second; and third, and last, those affecting the mental and nervous system.

TABLE I.

Those symptoms affecting the Generative Organs, and are local :

Nocturnal emissions, with or without erection or consciousness.
Pollutions accompanying or following defecation or stool.

Pollutions accompanying or following irritation on passing water.

Diurnal emissions.

Spermatic urine, or water containing animalcules.

Spasmodic or dull pains occasionally in the organs.

Premature emissions during or before coition or connection.

Contractions of the prepuce or foreskin.

Varicocele, or varicose veins in the testicles, generally the left.

Emissions, with erection upon slight excitement.

Emissions without erection on having lascivious thoughts.

Pimples on shoulder and forehead.

Priapism, or erection without any exciting cause.

Decrease of sexual desire or enjoyment.

Sanguineous emissions.

Diminution of the size of penis and testicles.

Want and imperfection of erectile power.

In answering all the above questions, you take them in their regular order, and answer them distinctly and positively. We have, then, the various general symptoms that may affect the different parts of the body, and that are all due to some over-indulgence of the sexual act, either by excessive venery or self-abuse ; and as many of these symptoms will be found to occur in, and to denote other forms of disease, still, if they are produced by the practice of self-abuse, they will be aggravated in degree, and will not yield to the ordinary treatment advised in such cases. But we must get at the root of the original disease. As an instance : In an otherwise healthy person, an attack of indigestion, originating in inattention to diet, will yield to gentle purgatives, tonics, and other well-known means; but if the symptoms of indigestion exist in consequence of the impairment of the nutritive functions by seminal losses, the ordinary remedies for such symptoms fail to produce their usual effect. Until the primary cause be removed, the trouble will be increased. We also find the same cause operating in disorders of the respiratory tract and the circulatory system, that when due to spermatorrhœa will not yield to the ordinary remedies; and it is a well-established fact that all the various functions of the human economy may be more or less deranged by this pernicious habit when long continued; hence we have the following symptoms:

TABLE II.

Those symptoms affecting and influencing the Muscular, Circulatory, Nutritive, and Respiratory Systems:

Excessive appetite.
Pain and heat in the stomach.
Unpleasant sensations before taking food, with disgust and heaviness afterward.
Weight in the stomach.
No desire for plain food.
Accelerated pulse.
Face flushed.
Regurgitations of food, with acid eructations.
Acrid heat in upper part of throat.

Secretions of liver and pancreas not normal, hence they will not assist the proper assimilation of the food.

Evolution of flatus.

Colicky pains in the bowels.

Cough, with difficulty of breathing.

Stomach and intestines distended with gas.

Flaccid or soft muscles.

Mucous secretions excessive.

Palpitation of the heart.

Apoplexy.

Liquid and unnatural stools, with constipation.

Hollow and sunken eyes.

Extreme sensibility to cold.

Alopecia, or loss of hair.

Inflamed eyes and eyelashes.

Indolence, or indisposition to work.

Lassitude and fatigue on slight exertion.

Rheumatism.

Shooting or fugitive pains in various parts of the body.

And so these general symptoms will pass on from one to another until they end in irredeemable debility. From these I now pass to those symptoms of the third class, that are so much worse in all their various characters and forms:

TABLE III.

Those symptoms affecting the Mental and Nervous Systems :

Blushing.

Want of confidence.

Avoidance of conversation.

Desire for solitude.
Restlessness.
Sighing.
Want of energy.
Uncertainty of tone of voice.
Want of purpose.
Aversion to society.
Depression of spirits.
Cowardice, or fear of solitude.
Listlessness, and inability to fix the attention.
Dimness of sight, with lachrymation.
Impairment of the hearing.
Vertigo, or dizziness.
Loss or impairment of memory.
Inequality of temper.
Peevishness.
Unable to fix the attention on any subject.
Desire for meditation.
Trembling of the hands or fingers.
Pain in the back of the head or spine.
Pain over the eyes.
Disturbed sleep, and unrefreshed in the morning.
Lascivious and peculiar dreams.

All these various symptoms will be found by the reader in the chapters on Nervous Debility and Spermatorrhœa; but I put them in this condensed tabular form that, in corresponding with me, they can so much more readily analyze their cases and give me a much better key to the proper treatment and management of their troubles, that I may guide and instruct them back to the paths of health and rectitude.

NOTES OF CASES.

As there are many of my readers that may be worried about "how to write to the doctor," and also to show them some of the many letters I am constantly receiving from my patients and their friends whom they have told to write to me, I will give these letters in full, only omitting their names or any allusion by which they may be recognized. All those who may write to me for sympathy and advice in their worry and distress may rest assured of the utmost privacy and secrecy.

These letters are selected from the hundreds on my file, that they may show the treatment and effects of the various diseases that I have spoken about in this book, and all the writers are aware and willing that I should publish them, that they may be the means of bringing "one more unfortunate" back to himself.

CASES.

"MY DEAR DOCTOR :—I have decided to write to you, from your well-known reputation and from the advice of a friend, in the hope that you may give me some relief from all my troubles. I will, therefore,

not trespass on your time. At the age of twenty-
one I came in possession of a handsome fortune, and,
with a large circle of friends, I had no troubles or
worry. My past life had been quiet and regular—not
addicted to intemperance nor masturbation—and I
was in as perfect a state of health as man could be.
At this time I desired to get married, but was prevented from doing so by my guardian from pecuniary
reasons ; and, in a fit of dissatisfaction, I formed several attachments of an illicit character, and, being
led away by my powerful sexual propensities, I indulged to excess—how much I need not confess;
but, from my twenty-first to twenty-fifth birthday, it
was almost my only occupation. At that time I felt
no diminution of power, but soon afterward my appetite for those indulgences began to lessen, and by
degrees my power also. I had neither desire nor
capability so often as before, and for a long period
would be indifferent to the matter; and then I noticed a constant feeling of lassitude and debility—
both bodily and mental—which unfitted me for any
active duty. I became dull, listless, peevish, appetite failed, and all the symptoms of dyspepsia set in,
for which I tried various remedies as well as consulting a physician. Though helped for a time, it was
only to return to the same condition, while my sexual powers were still worse, until I found myself almost impotent and sank into a miserable state of

despair and unhappiness. This has continued until the present time (now thirty-four years of age), when I come openly and candidly to you asking advice and assistance. I have no desire for sexual intercourse, and think I am losing my semen daily at stool and on passing water, and my organs are wasted and unable to perform the sexual act perhaps once in two or three months. Of my bodily sufferings I will not say much, but they are very severe, as my system is thoroughly debilitated and run down. I shall, therefore, await your answer with impatience, and will leave my case in your hands.

" Truly yours, ——."

[*Copy*].

When I received the above letter I wrote to him an encouraging answer, and advised him to come and see me at once. He did so, and, when he presented himself at the office, I was somewhat in despair, but told him I would do all that I could for him.

I therefore put him on appropriate general and local treatment, and also my Specific Pills. This treatment was kept up steadily for six months, when he returned home perfectly restored, and with my permission to get married.

A. R., a mechanic, put himself under treatment in 1877. Urine contained oxalate of lime and albumen and considerable mucous deposit. Violent headache at times, with great distress in stomach. Sometimes a difficulty in passing water, and attended with a scalding pain. Appetite very irregular, and slight pain in the lumbar region. Pain on the top of the head, restlessness at night, and frequent palpitation of the heart, and he looked pale and careworn. These symptoms continued for about one month, when they were very much abated; and, with the tonic pills and continued treatment, he was completely restored.

" DR. SMITH :—Desiring to place myself under your care, I did so, and with the most gratifying results. For the benefit of fellow-sufferers, I will give you a brief sketch of my case, which you have my full permission to publish. I was educated at a large boarding-school, where but little or no control was exercised over the morals of the boys, and I soon became a victim of the imprudent habits of youth. For some time I felt no bad effects, but ultimately, as I approached manhood, all the dreadful consequences of my self-abuse befell me. I dreaded society. My voice was harsh, my eyes weak and watery, my head heavy, memory and the sense of smell much impaired, appetite quite gone, and my strength of body and mind so destroyed that I was almost an imbecile. To add to this misery, the extreme emaciation into which I had fallen led to the suspicion that I was consumptive, and medical aid was sought by my friends. I was subjected to a series of questions, which worried and annoyed me, as of course I dared not confess the real cause of my destruction. I was thus falling into perdition, when I became acquainted with the efficacy of your medicines and *new method* of treating my complaint; and, thanks to your skill and judgment, I am again restored to health, my mind happy, and all my vigor returned— so much so, that I am now a happy and married man, and all owing to your skill and treatment.

"Yours, etc., ——."

"DEAR DOCTOR:—A friend of mine, who consulted you some months ago, Mr. ——, has strongly advised me to apply to you. I will state my case candidly, in hope you may do me some good. I enclose your fee, $5, and will give you my condition as candidly as possible.

· "I am twenty-eight years of age; I commenced the practice of masturbation at school, when only eleven years old, and continued it up to about one year ago. In justice to myself, I must say, I was not aware of the enormity of the vice, nor did I ever consider its destructive effects. A year ago, I desired to get married, but when considering this step, I discovered that I was not fit for the married state;

in fact, I was impotent. I now consulted a good, kind-hearted man, our family physician. He strongly advised me to put away all thoughts of marriage, and put me under treatment; and had he lived, I feel confident he would have cured me; unfortunately he died, and I then consulted various practitioners, sometimes obtaining transient benefit, but no lasting result. I have taken phosphorus, iron, strychnine, and belladonna, and various other drugs, all with partial effect; but I now know that I am losing semen involuntarily. I have taken proper means to ascertain this fact, and knowing such to be the case, I come to you, saying I have daily and nightly emissions, without sensations, and am impotent. Can you cure me? I am in your hands.

"Faithfully, B. E. D."

I wrote to this gentleman, telling him a personal interview was advisable, and when he came I discovered that the valves, which should close the seminal ducts, were paralyzed and open. He did not need much medicine, but required the most energetic treatment to the mouth of these seminal ducts, and with proper treatment the result was immediate and marked, and I was eventually able to tell him that he might marry as soon as he pleased, giving him some advice about it, and so he wrote me as follows:

"I am now three months married; have carefully

carried out your instructions, and I do not suffer any inconvenience. I have taken the pills with great advantage to my general health; yet I am satisfied all the medicine in the world, judging from my past experience, would never have effected a radical cure without the applications you used.

"That I am no longer impotent, you may judge when I tell you, on medical authority, that my wife is now carrying. If you approve of it I will continue the tonic pills, and send you P. O. order for a supply of them. Yours in trust, B. E. D."

A. B., a planter, aged twenty-nine, whilst at school, had been led into the habit of self-pollution, which he had continued for some years. Having become aware of the sin and danger of the act, he discontinued it; but on attempting to gratify his passions by the natural means, he found to his dismay that he was totally unable to accomplish the sexual act. The penis was incapable of firm and vigorous erection, and a discharge took place before an entrance could be effected into the female organ. His general health seemed good and his frame robust; this I attributed to his pursuits, which obliged him to be many hours daily in the open air. I would not undertake the case until he had given me a solemn promise never again to indulge in the vile practice of onansim. He was under treatment for ten weeks, during which he had the pleasure of observing a rapid improvement, and he was discharged cured. Shortly afterward he wrote me the following letter:

"NEW ORLEANS.

"MY DEAR SIR:—I am happy to inform you, that your treatment has continued successful, although when I applied to you I scarcely expected any great benefit; for, as I wrote you at the time, I had been suffering some years with debility and moroseness which my friends believed constitutional, but which I knew in my own mind arose from self-pollution. I heard of you from a gentleman who had been under your care, and was induced by him to consult you. I am grateful to say that I now feel fully restored to health and strength, never better in my life.   Yours sincerely,   A. B."

Mr. H., of Carroll County, Maryland, aged fifty-seven, called June, 1879, to be treated for impotency. As he had just married his second wife and manifested great anxiety concerning his weakened powers, I advised him to try the benefits of the Faradic current through the genito-sacral plexus. His occupation being such as to require almost constant travel, he was unable to follow my orders, so I supplied him with my Nerve Tonic Pills and full directions, with the result of a most marked improvement in his pro-creative powers, so that after a few weeks' continual use of the remedy he reported himself "well able to enjoy the sexual congress. . . . ."

.This extract is from a gentleman living in the South, who had indulged himself unrestrained until he had become completely exhausted and powerless. While in this condition he was strongly advised by his friends to marry and was himself extremely anxious to enter the marriage state with a young lady to whom he was ardently attached, but his condition compelled him to decline. He wrote to me as follows:

". . . . Sexual union is scarcely possible at all. At times I have imperfect indications of power, but they never come when I will them and they disappear in spite of all my efforts to perpetuate them. Oh! how mortified I have been at my vain attempts with females lately, and how wretched I have felt at the thought that it must always be so! Doctor, I can not live in this way. I don't care to do so-—and then in regard to the proposed marriage, what shall I do, what shall I say, how can I possibly excuse myself? This is misery indeed, and, if possible, make me a man again. . . . ."

I at once put him upon the proper treatment, and in the course of six weeks we saw such a decided improvement that he could complete his arrangements for being married, which occurred three months after, and shortly after he wrote me the following letter:

" MY DEAR DOCTOR :—All my fears are dissipated, and I can look forward to the future without fear and feel myself as capable in every way as I could desire. No failures have occurred, nor have I any reason to dread them in the future. In fact, I need a little restraint now, as your medicine has a truly magical effect, and I could not believe it did I not see its effects on my own person. But for you I should have been a miserable wretch not fit to live, and now am a happy, healthy husband.

"With best regards, etc., ———."

G. J. H., a watchmaker, aged twenty-seven, living in this city, came to me suffering with a constant loss of the generative fluid, occurring almost nightly and sometimes in the day, and attended with utter impotence. He felt, he said, as if his life-blood was draining away from him, each discharge being attended with the most terrible palpitations of the heart and deadly fainting fits; he was also troubled with a nasty, hollow cough. In this case the strengthening, astringent, and tonic properties of my medicines were well marked. Not only were the emissions checked and the due and wholesome vigor of the frame restored, but the cough and other physical symptoms were also arrested. When I saw him last, he told me with many expressions of gratitude he was enjoying better health than he had done for years before.

A gentleman called at my office about two years ago, complaining that having been married seven years he did not have any children. His age was thirty-three and that of his wife twenty-six, and she was quite healthy. I found that, although he had led a free life, his virile powers were undiminished, and that he was in the habit of having connection several times a week. On examining his urine by means of the microscope, I detected the spermatozoa, which proved a waste of semen, and they were not in a healthy condition. He admitted, though not without reluctance, that in his youth he had practiced onanism. I at once advised and treated him accordingly, particularly by my *new method*, and was successful beyond my own hopes, as he wrote me his wife was *enciente.*

"DR. SMITH :—Knowing your high professional standing and that I can consult you freely, I write respecting my present unfortunate condition. I am one who through ignorance—fatal ignorance—has acted against the laws of man and nature and injured myself, I fear irrevocably, by indulging in the odious practice of self-pollution. Would to Heaven that

some mentor had warned me in time of the consequences of my sin !

"I am now twenty-five and partner in a large firm.
I was only sixteen or seventeen when I first commenced this habit of self-pollution and have continued
it to a very recent period every week. I now feel a
heavy, dragging pain in the left testicle, which hangs
rather lower than the other. The penis seems small
and shriveled, and I frequently have emissions at
night. My water is quite clear and apparently
healthy. I therefore do not think there is any loss
of semen in that manner; but there is sometimes a
slimy discharge at stool, especially when I am bound
in my bowels, which is frequently the case. I find
myself very weak, and often have pains in my back.
I am very anxious to marry, but know that my generative organs are too feeble for coition, and had I
married in my present state, should have been miserable for life. I also fear that my mental faculties are
greatly impaired, as my memory is very bad and my
nerves unsteady. I frequently suffer from headache;
I feel drowsy and low-spirited, and my voice is husky
and not so strong nor so clear as formerly.

"I think I have now given you all my symptoms, and
shall send you a bottle of my urine for examination,
though there does not seem anything wrong about it
to me; it is quite clear and natural, though when I
pass it, the penis sometimes feels hot and inflamed at

the end. I enclose your fee and hope you will give me encouragement and promise success.

"Yours truly, ——."

Six weeks after I heard from him again, of which letter I give a short extract:

" All the urgent symptoms are much abated. Your letter was a great encouragement. Shall keep on, as I feel more energy and my spirits are first-rate; appetite decidedly improved. During the last three weeks had only one nocturnal emission, and that was very slight, and no discharge when at stool."

Again, four weeks more have passed, and he writes me and I give my readers the important part:

" I consider myself thoroughly cured, and do not think I shall need any more medicine, unless you think better to continue it for a while longer; just as you say. I am to be married in about two months, and shall be glad to see you out this way."

S. J. W., time-keeper in a large manufactory in this city, applied for consultation, complaining of a pain in the left side, exceedingly poor appetite, and general debility. An examination revealed great nervous depression, functional disease of the heart, and some disorganization of the kidneys, rendering him liable to sudden death at any time. Having told him his danger, and the necessity for a change of occupation, and relaxation from business for a few weeks, I commenced treatment, which was entirely successful in its results. The character of the urine was changed in a few days, and the albuminous secretion stopped. An improvement in his general health began forthwith, and he returned to his duties in four weeks, a well man. In this case we have the complaint of so many men who in their youth have been addicted to *masturbation*, and in consequence his sexual organs and sensibilities were so irritable that sexual union was utterly impossible. He had a plentiful secretion of seminal fluid, but the slightest attempt at connection, or even thinking about it at times, brought on immediate emissions, so that he was in reality powerless, and he had always been so. He had taken, I believe, every *cordial* and *tonic* that was advertised, but all to no purpose, and scarcely a hope of relief seemed left, and in this condition he came to see me. I treated him first by my *new method* very thoroughly, by direct application to the mouths of the *seminal* ducts, with the effect of at once stopping the emissions; but still the attempts at connection brought them on too soon, so that the act could not be consummated. To relieve this, I put him on my *nerve tonic pills* to nutrify and tone the parts, and improve the seminal fluid, and with the most perfect success.

This trouble of *too quick emissions* is very common,

and is both annoying and hurtful, for it will surely bring on seminal emissions, but in all cases I have been eminently successful.

A gentleman, aged forty, unfortunately the victim of misplaced confidence, was attacked with syphilis. Not knowing the terrible consequences of neglect, he did nothing but bathe the parts with a lotion for about four weeks, when the chancre began to increase in size, and to cause him some uneasiness; he also began to notice some irritation of the throat, and sore places in his mouth, with itching in the palms of the hands. At this stage he applied to me for treatment, and I at once pronounced it a case of the secondary form. The virus of the disease had been absorbed into the circulation, and affected the whole system. This condition is much to be dreaded, as it can not be cured without several weeks' treatment. There was one thing in his favor: he had not been drugged with mercury, the common remedy of the hospitals, and of many private practitioners; so that I had not two diseases to contend with, the venereal and mercurial. I removed all the main symptoms by judicious treatment, in two months, but continued the treatment until he was entirely cured.

In another case, a gentleman had just got married, and about a fortnight after he was surprised and

shocked to find himself suffering from what he described as true gonorrhœa. Several years before he
had a severe form of gonorrhœa, which after three
months he succeeded in getting cured, and he had
not the slightest appearance of the disease when getting married. His wife was an apparently healthy
young woman, and he had not any reason to suspect
her fidelity. On the contrary, he knew her to be
virtuous, yet how was the discharge to be accounted
for? He evidently was suffering much mental anxiety
on the subject when he sought my advice, and this
was after making use of the same treatment as during his former attacks, but with no benefit, hence he
wrote to me, and I advised him to call. He had
taken large quantities of copaiba, cubebs, and other
drugs. I made a careful examination, particularly
of the urine, both chemically and with the microscope,
and then discovered that he was not suffering from
true gonorrhœa, but from a mild form of inflammation contracted from his wife, who was, at the time
of marriage, suffering from a not uncommon disease,
termed leucorrhœa, or whites. The purest and most
virtuous females are subject to this ; nay, even vir-
gins have it frequently. Having discovered the true
nature of his disease, I had no difficulty nor delay
in curing it, and at the same time relieving his mind
from all shadow of doubt about his wife, who, at my
suggestion, was successfully treated for her complaint.

I was consulted a short time ago by a gentleman aged thirty-five, in a state of great mental anxiety. He had in youth led a " fast life," and had repeatedly contracted both gonorrhœa and chancre—an infection of a much more serious nature. For this latter disease he had been treated with mercury in great abundance, and had been pronounced cured. Believing this to be the case, he had married and became the father of two children ; both of whom had lived only a very short time. Latterly he had felt his health materially declining, and was suffering somewhat from eruptions on his head, back, and chest, while an ulcer appeared to be forming at the back of the throat. His wife, too, was evidently affected in a very similar manner, with sore throat, copper-colored eruptions, baldness, and very severe nocturnal pains.

The wife, who had not been saturated with mercury, rapidly recovered ; but her husband suffered as much from the improper remedies used (by the former medical attendant) as from the disease itself. By a thorough course of treatment, I ultimately succeeded in curing him of the disease and eradicating the mineral poison.

A few months ago I was consulted by a young man, who was employed in a large manufacturing establishment, for a trouble that seemed to be a very obstinate gleet. It was at once evident that he had been a votary of self-abuse—indeed, he said he could hardly escape, as all his companions were more or less addicted to the habit. Some two or three years back he had read many of the books published on this subject, and became thoroughly alarmed and disgusted with the propensity. He at length had recourse to illicit connection, and thus contracted gonorrhœa, when, through unskillful treatment, gleet was the sequelæ. In this dilemma he consulted me.

Now, about this time his symptoms were complicated and many. He was extremely weak, losing flesh, headache, cold perspirations, eyesight much affected (especially the left eye), frequent dizziness (especially when stooping), emissions, pains in the shoulders and spinal column, urine thick, and passed frequently in small quantities.

The result of my treatment was a complete recovery of this patient ; and, being restored to health, he sought out all those young men whom he knew to be guilty of onanism, warned them of their danger, and induced them to apply to me for the necessary treatment. They all had the good sense to follow my instructions and treatment and were cured.

Thus, through the candor, conscientiousness, and moral courage of one young man, a number of others were rescued from vice, disease, and misery, and brought back to health and happiness.

Some time ago a patient came to me suffering from a recently-contracted gonorrhœa and complicated by a spermatorrhœa in its active form. By the advice of a friend he had procured an astringent injection, which, however, he discarded just in time to escape an attack of orchitis, or swelled testicle, in consequence of its injudicious application. The local swelling and inflammation were intense; the discharge not very painful, but dark-colored and occasionally streaked with blood. The nocturnal emissions were frequent and profuse, and were attended with great pain of some hours' duration.

My first object was, of course, to remove the gonorrhœa, which, aggravated as it had been by improper treatment, soon yielded to my specific medicine, accompanied by certain subsidiary remedies, and within ten days was entirely arrested. The nocturnal emissions, however, continued as abundant as ever; though no longer attended with such severe pain. These I also succeeded in preventing by my *new method* treatment, and finally I completed the cure, and, by regenerative treatment, restored his constitution.

My readers will judge that this case required very skillful and intelligent treatment, and would not have admitted the usual plan of dosing with all the vile medicines commonly employed, and, if so, would have left behind a gleet or stricture.

This case came to me from Bridgeport, Conn., on account of a severe inflammatory gonorrhœa that he had contracted from illicit intercourse, and had tried the various remedies with no avail. When he presented himself, I made an examination and found him also suffering from a phimosis or elongation of the foreskin, and so much so, that he had never seen the glans of his penis, as he could not pull the foreskin back; in consequence of this, his disease had extended to the corona of the glans. Before putting him on the proper treatment, I was compelled to perform a slight operation on the foreskin, which relieved that trouble at once, and enabled him, for the first time in his life, to see the glans or head. I then treated him for his clap, and am happy to say that he has entirely recovered and finds that he can enjoy the sexual act much more than formerly.

As an opposite condition to the case last noted, Mr. B. D. B., a carpenter of Brooklyn, came to me suffering from a large number of superficial ulcers on the glans, foreskin, and dorsum or back of the penis. These were chancroids, and he had been to several physicians who had treated him with the various remedies in vogue, as iodoform, etc., but he continued to get worse, and having retracted the foreskin, he was unable to return it—so from the swelling of the mucous membrane and the parts underneath, he had

caused partial strangulation and a *paraphimosis.*
Again I performed a trivial operation to loosen the
strangulation, and with appropriate local and general
treatment for his ulcers, I soon saw an improvement
that went on to a perfect cure.

I will close my illustrations of cases with one showing the effects of excessive intercourse, and shall leave
the subject in the hands of my readers, with the assurance that all their cases and letters shall have the
utmost privacy, and that with all their consultations
with me they may be assured of the most perfect
sympathy and success in treatment.

Mr. A., a young man who had never been guilty of
self-abuse, but who had married before reaching the
age of twenty, and though he had not indulged in any
promiscuous intercourse, still he had indulged with
his wife until his health was very seriously affected.
He complained of dizziness, determination of blood
to the head, dimness of sight, and a failing memory,
together with a rapid decline of strength. Conjugal
intercourse, which was for some time very imperfect,
had become almost impossible.

On microscopic examination, I found he was suffering from a form of passive spermatorrhœa, and ordered him to abstain from any further attempts, with a
powerful course of remedies, together with the appli-

cation of my "*new method*" and the internal administration of my "*Nerve Tonic Pills*," as the case required. As he was blessed with a strong and vigorous
constitution and faithfully carried out all my instructions, I am happy to say that in two months his virile
powers were completely restored, and for the future
I only cautioned him to use moderation and not
excess.

As soon as he was satisfied that he was on the correct road to health, he consulted me about his wife,
who had also suffered from their mutual excesses.
Her case presented many difficulties, but by careful
and personal treatment I was able to restore her, and
carefully admonished them both to guard against this
folly and serious complication in the future.

SPECIAL NOTICE TO PATIENTS AND READERS.

It is now many years since I commenced the practice of treating these special diseases of the GENERATIVE and NERVOUS SYSTEMS, and I can therefore offer my patients unusual advantages, being supplied very fully with all that is necessary to treat the cases successfully, both SURGICAL and MEDICAL, as described in this book.

Dr. E. D. SMITH, surgeon, can therefore be personally consulted—as I do not put any of my cases in the hands of assistants—at his office daily, No. 100 East Twenty-ninth Street, during the hours of eight in the morning until one, and five until eight in the evening.

It is always desirable in all cases that I should have, at least, one personal interview with my patients, *even with those living at a distance*, that I may make at once a positive diagnosis, as this will result in manifold advantages to all that are afflicted, and is far superior to a mere correspondence, as a single visit in most cases will enable me to make an instantaneous and accurate judgment, *and thus expedite the recovery*, as a more correct diagnosis of the disorders, and a better appreciation of the patient's constitution, can be arrived at, whilst an exact microscopic examination of the urine, when necessary, will render any mistake impossible.

Especially is this the case in those suffering from spermatorrhœa. Also, there are so many important questions affecting the patient that could not be put in this book, and that would be suggested by seeing the patient myself, and that would be forgotten or overlooked in a correspondence. And in the case of those suffering from any urethral discharge, be it venereal or not, whether produced by an impure connection or from any cause, it is eminently better for me to see and examine the discharge for a correct diagnosis, as well as the facility for urinary examination ; and these many advantages will well repay one in the rapidity and permanent cure of whatever disease he may be suffering from.

But to those living at a distance from the city, I would advise them to write carefully according to my instruction of symptoms in another part of this book, and that it is to their own interest that they should be as confidential and as minute in giving me very full details of all their symptoms, age, habits of living, etc., etc., that I can send them the proper and necessary remedies, to any address, or directed to be left till called for, at any railway station or express office, in a portable compass, carefully packed and free from any observation, and that they may be taken without confinement or any restraint whatever— only to follow directions.

Hence, to those desiring to consult me, either personally or by letter, they can rely on the following rules of my office :

First. That all cases applying to me will be strictly confidential and secure.

Second. I can always be seen at the hours above named at my office, and at any other time only by special appointment by letter.

Third. That I will never undertake a case unless I can guarantee a cure, or very materially alleviate the existing trouble or deformity.

Fourth. That all my preparations are prepared by myself in my *laboratory*, and that they are perfectly pure and reliable.

Fifth. That I will not advise nor undertake any surgical operation unless I consider it absolutely necessary for the health and happiness of the individual.

Lastly. I can always and at all times be seen personally by my patients, and they may be assured of perfect sympathy and skillful advice.

All communications should be accompanied by my usual fee in such cases—*five dollars*—which may be sent by cash, check, or post-office order. In all cases, this rule is positive, and all letters will be considered inviolable, and will be either returned to the writers, destroyed at the termination of each case, or kept secret for future reference. I have many patients whose cases I have conducted to a speedy and successful termination without a single interview, and only by means of letters fully describing all their symptoms; as for a long series of years most of my practice was conducted by correspondence only, distance being no hindrance or additional expense to invalids residing in the most distant and remote parts of the country. Patients may also write in any language that may suit them, English, French, Spanish, etc.; and if they decide to use an assumed name or initials, it does not make any difference to me. I only request them to always keep the same name or initials throughout all the correspondence, and each letter should also contain the address to which the writer wishes his letters sent or packages directed in plain and distinct writing.

To those living at a distance, I would advise them that they can send me a sample of their urine in a small two-ounce vial (flat, securely *corked* and *sealed*), and packed carefully in cotton or sawdust in a small box, which may be obtained from any druggist; and always send me the first urine that is passed in the morning on rising. The parcel should be addressed plainly (*charges paid*) to Dr. EDWIN D. SMITH, No. 100 East Twenty-ninth Street, corner Fourth Avenue, New York City.

"Dr. Edwin D. Smith, of this city, is one of the most successful and reliable practitioners in all that branch of his profession pertaining to disorders of the nervous system and the genito-urinary organs. Having held several State-professorships and enjoyed a very large hospital experience, combined with his very extensive and enlarging private practice, also the author of several works and inventor of many surgical instruments, he is fully qualified to take the most difficult cases, in which he has enjoyed the most unvarying success and the thanks and regards of scores of the best people in the city."

The above extract was taken from one of our large city papers, and I insert it here for the benefit of my readers.

I would again impress upon my readers and those contemplating seeking my advice, the absolute necessity of having such matters attended to at once, as I am frequently enabled to cut short these diseases at the very start, whereby the patient is relieved of so much suffering and worry of mind and body. Hence the necessity that, if possible, they should without delay seek a personal interview at my office, thereby saving immense time and expense in the treatment of their complaints.

DR. EDWIN D. SMITH, *Surgeon,*

No. 100 East Twenty-ninth Street,

Office Hours :   New York City.

8 A.M. to 1 P.M., and 5 to 8 P.M.

Woman : Her Diseases and Treatment,

WITH THEIR

CAUSES and CURE,

CLEARLY EXPLAINED.

————◂▸————

This work, designed for the benefit of the female sex, has been published, and it will be found interesting and instructive. I have realized the necessity of such a work since I introduced the present volume of "THE NEW METHOD," a work designed for *men only*, and as I have received many letters, asking me if I had such a work for the female sex, I determined to issue the present edition, that I might give the female sex a true knowledge of their organs, and the various diseases pertaining thereto, and to point out to them a guide to health, that their lives may not be rendered miserable and unhappy. 200 pages, illustrated, and handsomely bound. Price One Dollar. Address,

DR. EDWIN D. SMITH,

100 East 29th St., New York,

CORNER FOURTH AVENUE.